Sustainable Economy and Ecotechnology

IEREK Interdisciplinary Series for Sustainable Development

The "Sustainable Economy and Ecotechnology" (SEE) book series acts as a timely initiative within the modern academic sphere, fostering a strong connection between technology, research, policy making, and a successful practical implementation for a sustainable future. In an age marked by increasingly urgent environmental challenges and by the critical importance of sustainable development. The series editor is dedicated to steering positive efforts to provide a strong platform for the dissemination of high-quality research focusing on sustainable business, technology, and ecological management and conservation. Through the contributions of experts in the field, "SEE" endeavors to create a transition toward sustainable economic models and technologically driven ecological awareness. The "SEE" series welcomes submissions of selected papers from conference proceedings, monographs, textbooks, and edited volumes, aiming to cultivate a comprehensive understanding of sustainable development by shining a spotlight on a diverse array of topics that collectively contribute to the advancement of a Sustainable Economy and Eco-technology. The scope of this series includes, but is not limited to: Sustainable business practices and management strategies and digitalization, circular economy principles and benefits, climate change impact on economy and business operation, renewable energy and its impact in driving economic growth, role of technology in ecological and environmental preservation, sustainable agriculture, sustainable economics, sustainability in digital business management, finance and investment, sustainable supply chain management: impacting marketing, finance, production, manufacturing and distribution, Implications of economic development on the environment, sustainable mobility and its economic impacts, sustainable marketing strategies and its socio-economic implications, sustainable urban and rural development impacts on economy and more topics related to business investment, economy, environment, and technology.

Rahul S. Mor · Mamta Gupta · Shilpa Sindhu ·
Sarita Chaudhary
Editors

Sustainable Business Management, Innovation and Technology

International Conference Proceeding on Sustainable Business Management, Innovation and Technology

Springer

Editors
Rahul S. Mor
School of Professional Business Education
Faculty of Business and Law
University of Northampton
Northampton, UK

Shilpa Sindhu
Maharaja Surajmal Institute
GGSIP University
New Delhi, Delhi, India

Mamta Gupta
Maharaja Surajmal Institute
GGSIP University
New Delhi, Delhi, India

Sarita Chaudhary
Maharaja Surajmal Institute
GGSIP University
New Delhi, Delhi, India

ISSN 3005-1274 ISSN 3005-1282 (electronic)
Sustainable Economy and Ecotechnology
ISBN 978-3-031-98140-1 ISBN 978-3-031-98141-8 (eBook)
https://doi.org/10.1007/978-3-031-98141-8

This Springer imprint is published by the registered company Springer Nature Switzerland AG
The registered company address is: Gewerbestrasse 11, 6330 Cham, Switzerland

About This Book

The Corporate Environment has undergone a paradigm shift in favor of sustainability in recent years. Businesses are under pressure to adopt sustainable practices. This book aims to explore the intersection of sustainable business management, innovation, and technology, providing insights into how organizations can thrive in a rapidly changing world while making positive contributions to the environment and society. It will empower businesses to embrace sustainability as a core principle and drive positive change in the world.

"Sustainable Business Management, Innovation, and Technology" offers priceless advice for embracing sustainability as a catalyst for growth and resilience in the twenty-first-century economy, regardless of your role: corporate leader, entrepreneur, legislator, or student.

The book consists of high-quality peer-reviewed research papers presented at the International Conference on *Sustainable Business Management, Innovation and Technology*. The book is meant for scholars, students, business professionals, policymakers, and entrepreneurs who are interested in learning how innovation, technology, and sustainable business management come together. It offers research papers, theoretical frameworks, and practical insights, making it an invaluable resource for novices and specialists alike.

The book gives readers the skills and information they need to successfully traverse the challenges of sustainable business management and prosper in a constantly shifting business environment through its blend of academic frameworks and real-world experiences.

As they study the foundational ideas of sustainable company management, readers will see how economic, environmental, and social factors are incorporated into organisational decision-making processes.

The book shows how companies can not only reduce their environmental effect but also stimulate innovation and long-term profitability through perceptive analysis and real-world case studies. By showcasing innovative approaches and technological solutions, this book will empower businesses to embrace sustainability as a core principle and drive meaningful change towards a more sustainable future.

The book delves into research articles spanning four distinct tracks and their respective sub-tracks, providing a comprehensive exploration of various dimensions of sustainable business management.

Track 1: Transforming Business Management for Sustainable Development

The chosen papers under this track provide a basis for comprehending and navigating the complex terrain of sustainable development within the field of business management by presenting a variety of viewpoints and insights into the multifaceted journey of integrating sustainability principles into business practices.

Track 2: Legal Aspects of Sustainable Development

This track focuses on research articles concerning the legal aspects of sustainable development. Within this thematic area, contributors delve into the intricate intersections between law, policy, and sustainable practices, offering critical insights into the legal frameworks and regulations that shape and govern sustainable development initiatives. Through rigorous analysis, the articles shed light on the evolving role of law in promoting environmental protection, social equity, and economic viability, thus providing a comprehensive understanding of the legal dimensions of sustainable development.

Track 3: Technology and Innovation

Continuing its exploration, this track covers a wide range of subjects, illustrating how innovation and technology meet sustainable development. Readers can learn important lessons about the transformative power of technical breakthroughs in promoting sustainability in a variety of sectors and businesses by reading these articles.

Track 4: Advances in Entrepreneurship and Economic Growth

The book proceeds with a collection of invited research papers spanning various subtracks under the overarching theme of "Advances in Entrepreneurship and Economic

Growth". The selective contributions explore various aspects of economic development and entrepreneurship, providing insightful information on cutting-edge tactics, new developments, and optimal practices that promote long-term, steady economic progress. These articles highlight the vital role that entrepreneurship, innovation, and sustainable development will play in determining the future of the global economy through their in-depth study and case studies.

Contents

A Performance Analysis of Random Permutation and Dimensionality Reduction Based Methods for Multimodal Cancelable Biometric

Onkar Singh, Ajay Jaiswal, and Naveen Kumar

Abstract Cancelable biometrics methods can be unimodal or multimodal. Unimodal cancelable biometric methods use single biometric traits, whereas multimodal methods use more than one. Unimodal methods are less expensive and require less cost and storage than multimodal methods, but multimodal methods are more accurate and secure. This paper applies existing random permutation and dimensionality reduction-based methods template generation methods for generating multimodal biometrics cancelable templates. The fusion of Georgia Tech face and CASIA-IrisV1 biometrics creates the multimodal biometric database. The performance analysis shows that these unimodal methods can be applied to multimodal biometric applications and improve/retain recognition accuracy. The templates generated using the fusions are also irreversible and immune to different impostors' attacks as evaluated using the Unified Average Changing Intensity (UACI) and Number of Pixel Change Rate (NPCR). If the templates are compromised, they can be cancelled and replaced with new ones that do not correlate with the old ones.

Keywords Random permutation · Cancelable biometrics · Multimodal biometrics · Template protection · Privacy · Security

O. Singh (✉)
Shaheed Sukhdev College of Business Studies, University of Delhi, Delhi, India
e-mail: onkarsingh@sscbsdu.ac.in

O. Singh · N. Kumar
School of Open Learning, University of Delhi, Delhi, India
e-mail: nkumar@cs.du.ac.in

A. Jaiswal
Department of Computer Science, Faculty of Mathematical Sciences, University of Delhi, Delhi, India
e-mail: ajayjaiswal@col.du.ac.in

R. S. Mor et al. (eds.), *Sustainable Business Management, Innovation and Technology*, Sustainable Economy and Ecotechnology, https://doi.org/10.1007/978-3-031-98141-8_1

1 Introduction

There are several ways to authenticate users and give secure access to IoT-based and other smart devices. These authentication methods are based on three factors: (a) what you know, (b) what you have, and (c) who you are (Ouda, 2016). The methods of the first category rely on the knowledge a person possesses to qualify as a genuine user, such as passwords and personal identification numbers (PINs). The second category of methods necessitates physically possessing an authenticated device, such as a smart card or token. And the methods belonging to the third category, such as biometric-based methods, are based on the person's physical characteristics. These characteristics include face, iris, ear, fingerprint, gait, voice, hand geometry, etc. (Manisha & Kumar, 2020a). The biometric-based user authentication methods are considered more reliable than the other two categories as they differentiate between authorized users and unauthorized users containing stolen passwords/PINs/cards, etc. The reason for the reliability is that a person's biometrics cannot be shared, stolen, or forgotten. The main advantage of the user authentication methods of the first two categories is that the passwords, PINs, smart cards, etc., can be cancelled and renewed if stolen. On the contrary, the biometric details, once stolen, are stolen forever due to their permanent association with the person. Thus, employing biometrics for user authentication prompts worries regarding user privacy and safeguarding biometric information. The cancelable biometric scheme is suggested in the literature to resolve these issues. The cancelable biometrics is defined as an intentional, repeatable distortion of the biometric signal based on a chosen transform (Ratha et al., 2001). It involves a deliberate and consistent distortion of a biometric signal using a selected irreversible transformation method. In cancelable biometrics, the biometric data transformed in an irreversible domain is known as *template*. If stolen or forgotten, the template can be cancelled and renewed, such as passwords. A cancelable biometric method has four requirements: (a) *non-invertibility* of templates, (b) revocation and *renewability* of the stolen templates, (c) *unlinkability* between stolen and renewed templates, and (d) good recognition *performance* (Manisha & Kumar, 2020a).

Since we consider each pixel a separate feature, the problem of *curse of dimensionality* arises because of high dimensions in the biometric images. Various dimensionality reduction techniques, such as principal component analysis (PCA), linear discriminant analysis (LDA), fisher's linear discriminant analysis (FLDA), locality preserving projection (LPP), etc., are used to deal with this problem in the field of pattern recognition (Kumar & Rawat, 2020). Some of these techniques are used in cancelable biometrics to generate cancelable templates, such as random permutation principal component analysis (RP-PCA) uses PCA (Kumar et al., 2018), random permutation-based linear discriminant analysis (RPLDA) uses LDA (Punithavathi & Geetha, 2021), random permutation-based locality preserving projection (RP-LPP) uses LPP (Kumar & Rawat, 2020). These techniques generate irreversible templates from biometric images while reducing dimensions simultaneously. Hence, the space requirement for storing templates is low, and it takes less time to match templates.

Cancelable biometric schemes can be unimodal (based on a single biometric) or multimodal (based on more than one biometric). Unimodal methods are less expensive and require less cost and storage than multimodal methods, but multimodal methods are more accurate and secure (Gawande & Golhar, 2018). Multimodal biometric systems solve the problems that come up with unimodal systems, such as intra-class variability, inter-class similarity, data quality, noise sensitivity, and non-universality (Paul & Gavrilova, 2012). A multimodal system uses the fusion of more than one biometric characteristic that helps improve recognition accuracy and provides enhanced security (Sumalatha et al., 2024). Multimodal biometric fusion integrates data from various biometric features to amplify strengths and mitigate shortcomings of the individual measurements (Gudavalli et al., 2012). The fusion can be performed at the sensor, feature, score, and decision level (Oloyede & Hancke, 2016).

Contribution

The RP-PCA (Kumar et al., 2018), RPLDA (Punithavathi & Geetha, 2021), and RP-LPP (Kumar & Rawat, 2020) effectively address conventional biometric systems' user privacy and biometric security issues. However, they are suggested for unimodal cancelable biometrics. Our contribution to the field of cancelable biometrics is as follows:

- The application of RP-PCA, RPLDA, and RP-LPP for multimodal biometric cancelable template generation.
- An analysis of the performance comparison between unimodal and multimodal biometrics using face and iris datasets.
- Non-invertibility analysis of the cryptic patterns using Correlation coefficient, Structure SIMilarity (SSIM), Unified Average Changing Intensities (UACI), and Number of Pixel Change Rate (NPCR).

The remaining paper consists of four additional sections. Section 2 reviews similar techniques, while Sect. 3 focuses on using RP-PCA, RPLDA, and RP-LPP in multimodal cancelable biometrics. Section 4 presents the experiments and their results, and Sect. 5 concludes the paper.

2 Literature Review

The cancelable biometric schemes that first use randomness to generate cancelable templates are GRAY-COMBO and BIN-COMBO (Zuo et al., 2008). The GRAY-COMBO is designed for grayscale images, where each row's grayscale values are circularly shifted to the right randomly using a specific offset. Randomly chosen rows are then combined using addition or multiplication to create cancelable iris templates. In contrast, the BIN-COMBO works for binary images and combines the rows using XOR (Zuo et al., 2008). PCA has been widely used for dimensionality reduction in the face and other biometric recognition. The work of PCA is used in RP-PCA (Kumar et al., 2018) for cancelable biometrics. In this work, the biometric images are first converted into cryptic patterns by applying random permutations. This random permutation is applied using person-specific random permutation matrices so that only templates of one person can be cancelled and renewed if compromised. A total scatter matrix from the cryptic patterns (y_i) is generated as follows:

$$S = \sum_{i=1}^{N} (y_i - \overline{y})(y_i - \overline{y})^T \tag{1}$$

where $\overline{y} = \frac{1}{N}\sum_{i=1}^{N} y_i$ is the mean of all cryptic patterns and N is the total number of training images. The transform matrix $W \in R^{q \times d}$ is generated by choosing the first d eigenvectors from the eigendecomposition of S, where q is the dimension of column vector representing the cryptic patterns and $d << q$. The W converts each cryptic pattern into a non-invertible template. RPLDA also creates intermediate templates similar to the cryptic patterns in RP-PCA and applies LDA to generate cancelable templates with reduced dimensions. RP-LPP applies LPP to the cryptic patterns set to generate cancelable templates. The GRAY-COMBO and BIN-COMBO generate cancelable templates from the iris, whereas RP-PCA, RPLDA, and RP-LPP generate templates from the face, iris, and ear biometrics separately.

3 Application to Multimodal Biometrics

This paper extends the application of the existing cancelable biometric template generation methods RP-PCA, RPLDA, and RP-LPP to multimodal cancelable biometric template generation. The idea of the application is shown in Fig. 1.

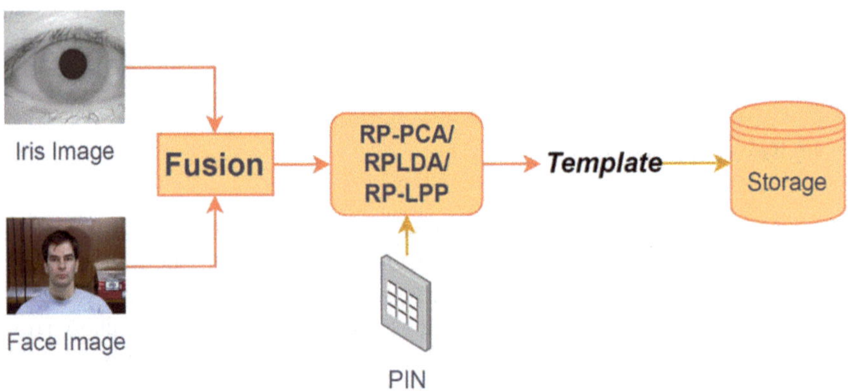

Fig. 1 Multimodal template generation using GT face (Nefian, 1999) and Casia-IrisV1 (Casia.V1)

Suppose we have N face and N iris training images for n persons, where each person represents a class. The biometric images of the face and iris are down sampled to $w \times h$, where w and h are the width and height of the image, respectively. We can also down sample the images of the face and iris datasets to different sizes, but we maintain the same sizes for simplicity. We represent the face and iris images as matrices $I_f \in R^{w \times h}$ and $I_e \in R^{w \times h}$, respectively. Now, the matrices $I_f \in R^{w \times h}$ and $I_e \in R^{w \times h}$ are converted into column vectors $x_f \in R^{q \times 1}$ and $x_e \in R^{q \times 1}$ using column concatenation, where $q = w \times h$. For example, the matrix A can be converted into column vector c as:

$$A = \begin{bmatrix} 1\,2\,3 \\ 4\,5\,6 \end{bmatrix} \Rightarrow c = \begin{bmatrix} 1 \\ 4 \\ 2 \\ 5 \\ 3 \\ 6 \end{bmatrix}$$

In this way, we represent a 2-dimensional matrix into a 1-dimensional column vector. Then, x_f and x_e column vectors are combined to perform the fusion of biometric images. The column vectors can be combined in several ways, such as by adding or multiplying them. Adding or multiplication results in the loss of their original features. Therefore, we combined both vectors through column concatenation, as we can retain their original features this way. Therefore, a new column vector $y \in R^{2q \times 1}$ represent the fusion of $x_f \in R^{q \times 1}$ and $x_e \in R^{q \times 1}$, i.e. x_f concatenated by x_e. Similarly, we fuse all pairs of face and iris images for each class in the training set, which is now represented through matrix Y as $Y = [\mathrm{y}_1 \mathrm{y}_2 \ldots \mathrm{y}_\mathrm{N}] \in R^{q \times N}$. If the size of the face and iris image is different, then we take the sum of the dimensions of both vectors, for example, if $x_f \in R^{p \times 1}$ and $x_e \in R^{q \times 1}$, then $y \in R^{r \times 1}$, where $r = p + q$. Therefore, to generate multimodal cancelable biometric templates using RP-PCA, RPLDA, and RP-LPP, we use this fusion of face and iris as input rather than a single biometric image.

4 Experiments and Results

In this section, we discuss the experimental setup and results. The experimental setup discussion includes the details of the biometric datasets used, whereas the result subsection analyses the templates' classification accuracy, non-invertibility, and security against the brute force attack.

Table 1 Details of biometric databases used

Database	Original dimension	Number of persons	Number of images	Images per person
GT face	640 × 480	50	750	15
CASIA.IrisV1	320 × 280	108	756	7

4.1 *Experimental Setup*

To illustrate the efficacy of the suggested methods, we conducted tests on a multi-modal dataset created using Georgia Tech (GT) Face (Nefian, 1999) and CASIA-IrisV1 (Casia.V1) datasets. Different scales, locations, and facial expressions illustrate the frontal and angled faces at Georgia Tech (Nefian, 1999) dataset. Before utilizing the template-generation approach, the color face images of GT were transformed to grayscale. According to the CASIA-IrisV1 (Casia.V1) database, the first session captured three images while the second session recorded four. Images from both sessions were pooled and converted from BMP to JPG format before experimentation. The details of biometric datasets are given in Table 1.

We used 7 images per person for 50 persons for both datasets to equate the datasets. Therefore, in the GT face dataset, each person's first 7 images out of 15 are used, and other 8 images are discarded, i.e., 350 images are used. In the CASIA-IrisV1 dataset, the iris images of the first 50 persons out of 108 are used for the experimentation. The images of both datasets were downsampled to 100 × 75.

4.2 *Result Discussion*

Each individual's biometric images are randomized using a seed value of 55 to ensure result stability and are subsequently divided into training and test sets for each database. We used two templates to train the classification model and five biometric images for testing. The K-nearest neighbours (K-NN) method was employed with a single neighbour to classify templates. Cross-validation is adopted to evaluate the model performance since it reduces the influence of randomness on test results.

4.2.1 Performance Analysis

The performance analysis shows the classification accuracy of all three methods, i.e. RP-PCA, RPLDA, and RP-LPP, on the Casia iris, GT face, and the fusion of face and iris. The reported results are shown in Table 2. The classification accuracies reported in the table can be visualized in Fig. 2. The classification accuracy of the templates generated using the fusion of iris and face image at just ten components is higher than the accuracy of the iris and face alone for RP-PCA and RP-LPP. For

RP-PCA, it is 95.60% for the fusion, whereas, for the iris and face, the accuracy is 91.20% and 90.40%, respectively. For RP-LPP, the classification accuracy for the fusion is 95.47%, but for the iris and face, it is 93.33% and 90.67%, respectively. However, the fusion accuracy is comparable to GT face in the case of RPLDA for 10 discriminants, where the fusion accuracy is 62.13%, and the accuracy of the face alone is 62.27%.

Table 2 Accuracy analysis

Method	No. of comp	2	4	6	8	10	12	14
RP-PCA	RP-Casia	34.00	62.67	75.07	83.87	91.20	93.60	95.87
	RP-Georgia	63.73	84.13	87.20	89.87	90.40	90.93	91.47
	RP-Cas + Georg	61.73	88.13	92.67	95.07	95.60	95.73	96.40
RP-LPP	RP-Casia	28.93	61.87	78.67	89.20	93.33	96.13	98.13
	RP-Georgia	49.87	78.00	87.60	90.00	90.67	91.20	91.87
	RP-Cas + Georg	42.80	80.80	90.67	94.00	95.47	97.20	97.20
RPLDA	RP-Casia	16.13	32.80	40.67	47.73	52.53	58.80	60.00
	RP-Georgia	19.47	36.93	48.40	56.93	62.27	66.67	69.73
	RP-Cas + Georg	8.80	23.47	38.93	55.73	62.13	68.00	71.73

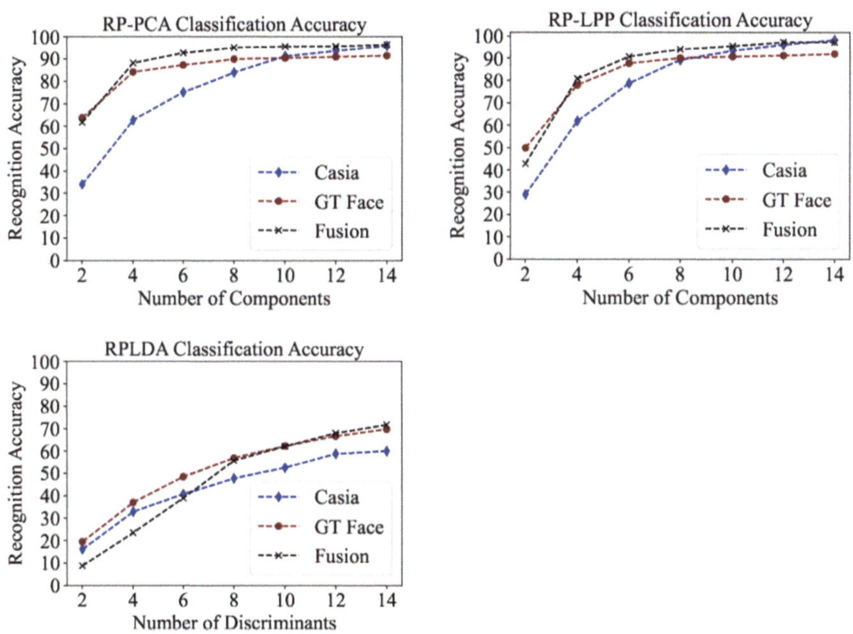

Fig. 2 Classification accuracy

For 14 components, the classification accuracy of the templates generated using fusion is higher for RP-PCA and RPLDA, whereas it is comparable with CASIA-IrisV1 for RP-LPP. Therefore, it is proved that the classification accuracy of the multimodal cancelable biometric templates generated through RP-PCA, RP-LPP, and RPLDA using the fusion of CASIA-IrisV1 and GT face is higher than that of a single biometric and sometimes comparable.

4.2.2 Non-Invertibility Analysis

The templates generated using the fusion of the iris and face are analysed for privacy and security. Literature suggests the use of the Correlation coefficient (C_r) and Structure SIMilarity (SSIM) to check the association and dissimilarity between the two images (Manisha & Kumar, 2020b). A C_r value of $+1$ or -1 indicates a positive or negative correlation; however, a value of 0 indicates no correlation. The C_r be computed as:

$$C_r(B,T) = \frac{\sum_i \sum_j (B_{ij} - \overline{B})(T_{ij} - \overline{T})}{\sqrt{\left(\sum_i \sum_j (B_{ij} - \overline{B})^2\right)\left(\sum_i \sum_j (T_{ij} - \overline{T})^2\right)}} \tag{2}$$

where B_{ij} and T_{ij} are biometric and template images, respectively; $\overline{B}$ or $\overline{T}$: the mean of B or T. The SSIM value of 1 represents the same images, and a value of 0 indicates no similarity. The SSIM can be computed as follows:

$$SSIM\,(B,T) = \frac{(2\mu_B\mu_T + C_1)(2\sigma_{BT} + C_2)}{\left(\mu_B^2 + \mu_T^2 + C_1\right)\left(\sigma_B^2 + \sigma_T^2 + C_2\right)} \tag{3}$$

where μ_i is the mean, σ_i is the variance, σ_{BT} is the covariance of B & T, and C_i is the constant. The Number of Pixel Change Rate (NPCR) measures the percentage of pixels changed between two images and is defined as follows:

$$NPCR(\%) = \frac{\sum_{i,j} D(i,j)}{W \times H} \times 100 \tag{4}$$

where $D(i,j) = 1$ if the corresponding pixels of two images differ and 0 otherwise. The Unified Average Changing Intensity (UACI) measures the average intensity variation between plain and encrypted images and can be computed as:

$$UACI(\%) = \frac{1}{W \times H} \times \left(\frac{\sum_{i,j} B(i,j) - T(i,j)}{255}\right) \times 100 \tag{5}$$

where 255 is the maximum intensity value of a pixel in the grayscale image. The cryptic pattern is generated by taking the dot product of the biometric image and

Table 3 Non-invertibility analysis

Method	C_r	SSIM	NPCR	UACI
Original	1	1	0	0
Cryptic pattern	– 0.0026	0.0117	99.5067	50.4312

person-specific random permutation matrix in RP-PCA and RP-LPP. In RPLDA, this cryptic pattern is referred to as an intermediate template. Here, we apply random permutation to the fusion of face and iris images to construct a cryptic pattern. The randomly selected fusion of original iris and face images is compared using (2), (3), (4), and (5) with its corresponding cryptic pattern, and the results are shown in Table 3. The ideal value reflects the comparison of identical images. Therefore, a sound cancelable biometric system should yield opposite results for these performance measures.

A C_r value of – 0.0026 and SSIM of 0.0117 are approaching zero, indicating that the fusion and its cryptic pattern are dissimilar and have no association. A UACI > 33.4635% and NPCR > 99.6094% indicate that an image encryption technique resists different impostors' attacks (Essaid et al., 2018). The obtained NPCR value of 99.5067, which is very close to 99.6094%, and UACI of 50.4312 between the fusion and cryptic pattern show that the cryptic patterns are immune to different impostors' attacks. However, the cryptic pattern can be reversed if the key is provided. The template is generated by extracting feature from the cryptic patterns using principal component analysis (PCA) in RP-PCA which reflect the coefficients corresponding to the selected eigenvectors, and these features are irreversible (Kumar et al., 2018). The features are extracted similarly using LPP in RP-LPP and LDA in RPLDA. The template is stored in the form of extracted features.

4.2.3 Diversity and Revocability Analysis

The absence of correlation between the randomly generated keys ensures that templates using different keys stay unrelated (Lee et al., 2018). This allows the usage of different keys across applications or systems to create uncorrelated templates from the same biometrics. It satisfies the diversity (unlinkability) criteria of the cancelable biometrics. Similar to the application of RP-PCA, RP-LPP, and RPLDA to unimodal cancelable biometrics, the templates generated through the iris and face fusion are also renewable in case of compromise by changing the person-specific key. The changed key generates the new random permutation matrix, resulting in a different cancelable template from the same fusion.

4.2.4 Security Analysis Through Brute-Force

Security analysis examines an adversary's ability to create an accurate cancelable template through any attack (Abdullahi et al., 2024). In a brute force attack, the attacker lacks knowledge of the transformation function, transformation parameters, original biometrics, or the transformed template (Abdullahi et al., 2024). Consequently, the adversary explores every potential combination to produce the altered template (Lee et al., 2021). To produce a template, the adversary must guess the cryptic pattern correctly. Randomly guessing an image requires up to $q!$ iterations, where q is the dimension of the column vector representing the cryptic pattern. As the dimension of the cryptic pattern is 7500 pixels, this process requires at least 7500! iterations. Estimating a random image of this high dimension demands centuries using brute force analysis, even for a supercomputer operating at 10^{17} floating-point operations per second (FLOPS) (Kumar et al., 2018). Therefore, a template generated from such a high-dimensional vector is secure from brute force attacks.

5 Conclusion

This paper applies the current cancelable biometric scheme, Random Permutation Principal Component Analysis (RP-PCA), Random Permutation-based Locality Preserving Projection (RP-LPP), and Random Permutation based Linear Discriminant Analysis (RPLDA) to generate cancelable templates from multimodal biometrics. The fusion of Georgia Tech face and CASIA-IrisV1 iris through column concatenation is the input, replacing a simple unimodal biometric image. The performance analysis reveals that the classification accuracy of the templates generated through RPPCA, RP-LPP, and RPLDA using the fusion of CASIA-IrisV1 and GT face is higher than that of a single biometric and sometimes comparable to it. The iris and face fusion templates meet the essential requirements for the ideal cancelable biometric system. The privacy concern is addressed since the templates created using the fusion are not invertible. The templates are secure, as shown by using brute force analysis. They can be renewed by altering the person-specific key in case of a compromise. Therefore, these existing cancelable biometric template generation techniques based on random permutation and dimensionality reduction can be securely used to generate multimodal cancelable biometric templates.

Acknowledgements We are thankful to the following authors/academies for providing the following biometric databases, which were instrumental in conducting our research:

CASIA-IrisV1: Portions of the research in this paper use the CASIA-IrisV1 collected by the Chinese Academy of Science's Institute of Automation (CASIA). *Database Link*: http://biometrics.idealtest.org/

Georgia Tech Face: Dr. Ara V. Nefian, *Database Link*: https://www.anefian.com/research/face_reco.htm

Author Contributions Onkar Singh and Ajay Jaiswal contributed to the study's conception and design. Onkar Singh wrote the manuscript's first draft, and all authors reviewed and commented on previous versions. All authors read and approved the final manuscript.

Funding No funding

Declarations

Conflict of interest/Competing interests The authors have no competing interests to declare relevant to this article's content.

References

Abdullahi, S. M., Sun, S., Wang, B., Wei, N., & Wang, H. (2024). Biometric template attacks andrecent protection mechanisms: A survey. *Information Fusion, 103*, Article 102144.

Casia.V1 (n.d.). Casia-IrisV1 database. http://biometrics.idealtest.org/

Essaid, M., Akharraz, I., Saaidi, A., & Mouhib, A. (2018). A new image encryption scheme based on confusion-diffusion using an enhanced skew tent map. In *Procedia Computer Science, 127, 539–548. Proceedings of the First International Conference on Intelligent Computing in Data Sciences*, ICDS2017.

Gawande, U., & Golhar, Y. (2018). Biometric security system: A rigorous review of unimodal and multimodal biometrics techniques. *International Journal of Biometrics, 10*, 142–175.

Gudavalli, M., Raju, S. V., Babu, A. V., & Kumar, D. S. (2012). Multimodal biometrics—Sources, architecture and fusion techniques: An overview. In *2012 International Symposium on Biometrics and Security Technologies*, pp. 27–34.

Kumar, N., & Rawat, M. (2020). Rp-lpp: A random permutation based locality preserving projection for cancelable biometric recognition. *Multimedia Tools and Applications, 79*, 2363–2381.

Kumar, N., Singh, S., & Kumar, A. (2018). Random permutation principal component analysis for cancelable biometric recognition. *Applied Intelligence, 48*, 2824–2836.

Lee, D. H., Lee, S. H., & Cho, N. I. (2018). Cancelable biometrics using noise embedding. In *2018 24th International Conference on Pattern Recognition (ICPR)*, pp. 3390–3395.

Lee, M. J., Teoh, A. B. J., Uhl, A., Liang, S.-N., & Jin, Z. (2021). A tokenless cancellable scheme for multimodal biometric systems. *Computers & Security, 108*, Article 102350.

Manisha, & Kumar, N., (2020b). On generating cancelable biometric template using reverse of boolean xor. In *2020 International Conference on Emerging Trends in Communication, Control and Computing (ICONC3)*, pp. 1–4.

Manisha, & Kumar, N. (2020a). Cancelable biometrics: A comprehensive survey. *Artificial Intelligence Review, 53*, 3403–3446.

Nefian, A. V. (1999). Georgia tech face database.

Oloyede, M. O., & Hancke, G. P. (2016). Unimodal and multimodal biometric sensing systems: A review. *IEEE Access, 4*, 7532–7555.

Ouda, A. (2016). A framework for next generation user authentication. In *2016 3rd MEC International Conference on Big Data and Smart City (ICBDSC)*, pp. 1–4.

Paul, P. P., & Gavrilova, M. (2012). Multimodal cancelable biometrics. In *2012 IEEE 11th International Conference on Cognitive Informatics and Cognitive Computing*, pp 43–49.

Punithavathi, P., & Geetha, S. (2021). Random permutation-based linear discriminant analysis for cancelable biometric recognition. In S. M. Thampi, E. Gelenbe, M. Atiquzzaman, V. Chaudhary, & K.-C. Li (Eds.), *Advances in Computing and Network Communications* (pp. 593–603). Springer Singapore.

Ratha, N. K., Connell, J. H., & Bolle, R. M. (2001). Enhancing security and privacy in biometrics-based authentication systems. *IBM Systems Journal, 40*, 614–634.

Sumalatha, U., Prakasha, K. K., Prabhu, S., & Nayak, V. C. (2024). A comprehensive review of unimodal and multimodal fingerprint biometric authentication systems: Fusion, attacks, and template protection. *IEEE Access, 12*, 64300–64334.

Zuo, J., Ratha, N. K., & Connell, J. H. (2008). Cancelable iris biometric. In *2008 19th International Conference on Pattern Recognition*, pp 1–4.

AI and Ethics in Research: A Bibliometric Insight

Amit Kumar and Rashmi Jha

Abstract The implication of Artificial Intelligence in research has significantly enhanced the efficiency of scientific research. However, it has also led to increase in academic misconduct, including data falsification and text plagiarism. The unethical use of AI in research publications is a threat for the research integrity and has becomes a hindrance to scientific progress. The present study aims to consolidate the existing literature on this subject area by analyzing publication trends over the time, most prolific authors, institutions, countries, collaboration patterns, document types, and citation patterns. To fulfil the objectives of the study a bibliometric analysis was conducted on 972 publications that were listed in the Scopus database. The first publication related to AI's role in identifying plagiarism dates back to 1980 and it was presented at a conference. The highest number of publications were recorded in the year 2023 accounting the highest number of citations. V. Wiwanitkit was identified as the most prolific author in this field. This study area mainly covered in the subject domains of computer science, medicine, and social sciences, with articles and conference papers as the predominant document types. The United States was the leading contributor to this area of research. The outcomes of this study may be useful to all stakeholders with an interest in this research domain.

Keywords Artificial intelligence · Research ethics · Bibliometrics · Academic integrity

---

A. Kumar (✉)
Central Library, Gurugram University, Gurugram, India
e-mail: amitkumar@gurugramuniversity.ac.in

R. Jha
Department of Engineering and Technology—CSE, Gurugram University, Gurugram, India
e-mail: rashmijha@gurugramuniversity.ac.in

R. S. Mor et al. (eds.), *Sustainable Business Management, Innovation and Technology*, Sustainable Economy and Ecotechnology, https://doi.org/10.1007/978-3-031-98141-8_2

1 Introduction

The rapid growth in the field of artificial intelligence (AI) has changed the landscape of research. AI has enabled researchers to analyze vast amount of data, identify patterns, and make predictions with high accuracy (Aguilera Cora et al., 2024; Jordan & Mitchell, 2015). However, with the increasing reliance on AI in research has also raised alarms for the research ethics. One of the main concerns about the use of AI in research is the possibility of bias and discrimination. AI algorithms have the capability to propagate existing biases, particularly when it trained on biased data, leading to unfair and discriminatory outcomes (Ali et al., 2020; Caliskan et al., 2017). A study by Ali et al. (2020) discovered that deep learning models used in the Citation recommendations system can unfairly influence the selection of articles for publication. This shows the importance of researchers being aware of potential biases in AI algorithms and taking steps to minimize them.

Another concern is the lack of transparency and accountability in AI decision-making processes as AI becomes more autonomous. This makes it difficult to identify and correct errors or biases (Ampatzoglou et al., 2019; Doshi-Velez & Kim, 2017). This lack of transparency can undermine confidence in research results and undermine the integrity of the research process.

Despite these concerns, AI also provides opportunities to improve research policy. For example, AI can help identify and reduce threats to validity in research studies, such as selection bias and confounding variables (Ampatzglou et al., 2019; Mehrabi et al., 2019) AI. The present paper aims to provide insight into the current state of AI and research ethics. The present study will review the existing literature on the use of AI in research, including its benefits and challenges and identify areas where AI is used concerning research ethics.

2 Objectives

The prime objective of this study is to explore the pattern of literature growth on AI vis-à-vis research ethics. The specific objectives of the study are:

1. To investigate the year-wise distribution of publications on the topic;
2. To explore the type of documents under which the literature is published;
3. To identify the most prolific authors, institutes, and countries publishing in this field;
4. To explore the collaboration network among the different countries;
5. To know the most preferred sources of publications;
6. To explore the most frequently used keywords and visualize the co-occurrence among them.

3 Methodology

The present study used a bibliometric analysis to evaluate literature growth on AI vis-à-vis research ethics. To full the objectives of the study the data was collected from the Scopus with help of the search string " ((TITLE-ABS(AI) OR TITLE-ABS-KEY("Artificial intelligence")) AND (TITLE-ABS-KEY("Research Ethics") OR TITLE-ABS-KEY(Plagiarism) OR TITLE-ABS-KEY(research misconduct) OR TITLE-ABS-KEY("publi* ethics")) AND (LIMIT-TO (LANGUAGE, "English")))". Using the string above, the data was exported in CSV format and analyzed by using MS Word and VOS viewer.

4 Data Analysis

4.1 Number of Publications Per Year

The first paper in which the use of AI in the detection of plagiarism was discussed was published in 1980. Table 1 shows a fluctuating trend in publications and citations in the last 44 years. The highest number of publications (369), which is nearly 38% of the total publications with a growth rate of 459.09%. The highest number of citations (6929) were also recorded in the year 2023, which is nearly 53 of the total citations. The average citations per publication (ACPP) was 18.78 in 2023. The field has experienced significant growth in recent years, with an increase in both publications and citations.

4.2 Document Types

A total of 972 publications have been published on this topic which received 13,001 with an average of 13.38 citations per paper. Table 2 shows that these publications were published in form of 12 different document types, with a total of 972 publications. Article was the most common type of document accounting for 384 (39.51%) publications. Followed by Conference Papers and Editorials with 214 and 87 publications respectively. These top three document types account for nearly 70% of total publications. Review and Notes received the highest average citation per paper citations, with an average of 35.59 and 24.48 per publication, respectively.

Table 1 Year-wise publications

Year	NP	% of TNP	AGR (%)	CR	% of TC	ACPP
2024	287	29.53	– 22.22	1000	7.69	3.48
2023	369	37.96	459.09	6929	53.30	18.78
2022	66	6.79	22.22	632	4.86	9.58
2021	54	5.56	8.00	1019	7.84	18.87
2020	50	5.14	117.39	909	6.99	18.18
2019	23	2.37	35.29	784	6.03	34.09
2018	17	1.75	41.67	541	4.16	31.82
2017	12	1.23	– 25.00	113	0.87	9.42
2016	16	1.65	100.00	148	1.14	9.25
2015	8	0.82	– 20.00	54	0.42	6.75
2014	10	1.03	11.11	144	1.11	14.40
2013	9	0.93	125.00	208	1.60	23.11
2012	4	0.41	– 42.86	31	0.24	7.75
2011	7	0.72	16.67	41	0.32	5.86
2010	6	0.62	– 45.45	64	0.49	10.67
2009	11	1.13	37.50	137	1.05	12.45
2008	8	0.82	700.00	129	0.99	16.13
2007	1	0.10	– 66.67	0	0.00	0.00
2006	3	0.31	– 25.00	35	0.27	11.67
2005	4	0.41	300.00	16	0.12	4.00
2004	1	0.10	0.00	16	0.12	16.00
2003	1	0.10	– 50.00	12	0.09	12.00
2002	2	0.21	100.00	27	0.21	13.50
2000	1	0.10	0.00	6	0.05	6.00
1987	1	0.10	0.00	5	0.04	5.00
1980	1	0.10	– 99.90	1	0.01	1.00
Total	972	100.00		13,001	100.00	13.38

NP = Number of Publications, *TNP* Total Number of Publications, *AGR* = Average Growth Rate, *CR* = Citations Received, *TCR* = Total Citation Received, *ACPP* = Average Citations Received

4.3 Most Prolific Authors

Figure 1 illustrates most prolific authors in the field of research ethics concerning with the use of AI in research activities. The figure reveals that 14 authors have at least 3 publications, with a total of 55 publications. Wiwanitkit, V. was the most prolific author, with 10 publications, followed by Kleebayoon, A. with 9 publications. It was also noticed that the majority of authors (11 out of 14) have published 3–5 papers,

Table 2 Document type

Document type	NP	% of TNP	CR	ACPP
Article	384	39.51	6029	15.70
Conference paper	214	22.02	1043	4.87
Editorial	87	8.95	1761	20.24
Review	80	8.23	2847	35.59
Letter	49	5.04	138	2.82
Erratum	47	4.84	32	0.68
Note	44	4.53	1077	24.48
Book chapter	39	4.01	30	0.77
Conference review	19	1.95	0	0.00
Book	6	0.62	37	6.17
Short survey	2	0.21	5	2.50
Retracted	1	0.10	2	2.00
Total	972	100.00	13,001	13.38

NP = Number of Publications, *TNP* Total Number of Publications, *CR* = Citations Received, *ACPP* = Average Citations Received

indicating a moderate level of research output. The figure highlights the most prolific authors and potential areas of collaboration.

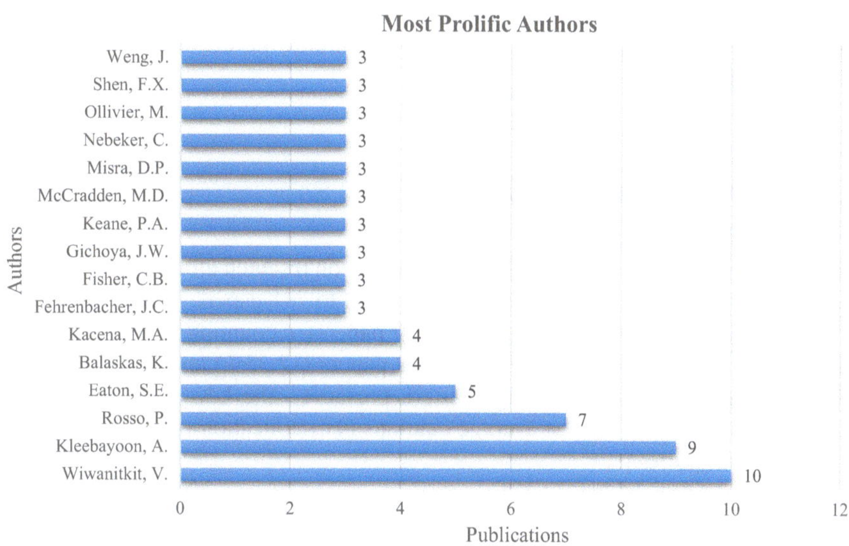

Fig. 1 Most prolific authors

Table 3 Most prolific institutes

Sr. No.	Affiliation	Country	Publications
1	University of Toronto	Canada	16
2	University College London	United Kingdom	14
3	University of Oxford	United Kingdom	13
4	Chandigarh University	India	12
5	Harvard Medical School	USA	11
6	University of Leeds	United Kingdom	10
7	Stanford University	USA	10
8	University of Calgary	Canada	8
9	The University of Manchester	United Kingdom	8
10	University of Washington	USA	8
11	Stanford University School of Medicine	USA	8

4.4 Most Prolific Institutes

Table 3 shows the most prolific institutes working in this field. The table reveals that 11 institutes have at least 8 publications, with a total of 108 publications. The University of Toronto has the highest number of publications (16 papers), followed by University College London and the University of Oxford with 14 and 13 publications respectively. Both the institutes (University College London and the University of Oxford) are locationed in the UK. The data reveals the strong position of UK institutions. Among the top 11 most prolific Institutes the USA and the UK both have 4 institutes each.

4.5 Most Productive Countries

Table 4 reveals a global landscape of publications on research ethics concerning AI. It was found that the top 10 countries contributed to 741 papers. The United States has the highest number of publications (239 papers) accounting for 24.59% of all publications. Followed by the UK with 120 papers (12.35%) and India 81 papers (8.33%). The data shows a strong Western presence. The UK, Australia and the United States have the highest average citations per publication (ACPP), indicating a high research impact.

Table 4 Most productive countries

Country	Documents	% of TP	CR	ACPP
United States	239	24.59	4186	17.51
United Kingdom	120	12.35	2550	21.25
India	81	8.33	591	7.30
China	60	6.17	383	6.38
Canada	57	5.86	633	11.11
Australia	47	4.84	839	17.85
Germany	45	4.63	486	10.80
Italy	39	4.01	613	15.72
Spain	27	2.78	259	9.59
France	26	2.67	225	8.65
Total	741	76.23	10,765	14.53

4.6 Country Collaboration

A total of 111 countries have contributed to literature in this field. Figure 2 illustrates the collaboration among the countries which have at least 5 publications. A total of 50 such countries were found that were grouped into 8 clusters. The USA, UK and India were the most productive countries. These countries have the highest number of publications and citations. These countries also have a large number of publications in collaboration. The analysis found eight groups of countries connected by 297 links, showing strong global interest in AI ethics research. This indicates a high level of international cooperation.

4.7 Most Preferred Sources

This literature work is published under 160 different sources. Table 5 shows the top 10 most preferred sources under which this literature is published. A total of 153 publications (15.74%) were published under these top 10 sources. BMJ Open and Lecture Notes are the top 2 sources in this field, with the highest number of papers. They include 33 and 32 publications, respectively. The data highlight a variety of sources, including journals, conference reports, and book collections. The existence of high-impact sources such as Nature and the American Journal of Bioethics indicates a multidisciplinary research approach.

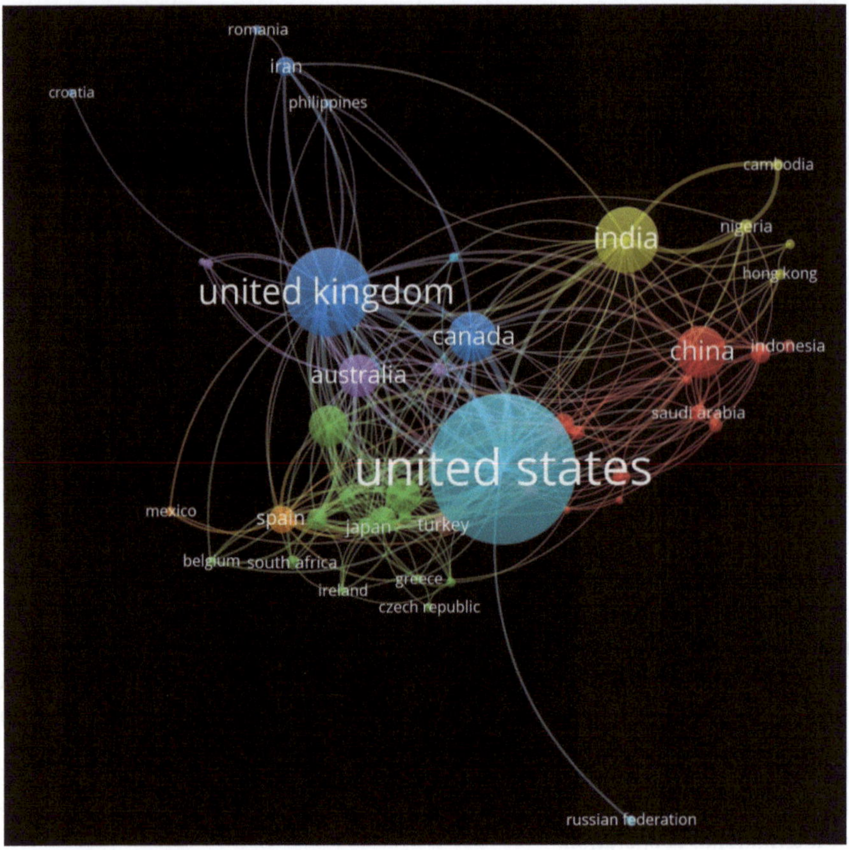

Fig. 2 Country collaboration

4.8 *Most Frequently Used Keywords and Their Co-Occurrence*

A total of 1,861 author keywords were used in the literature, with 99 author keywords used at least 4 times. These 99 keywords belong to 8 different clusters. This indicates a wide range of topics and themes. The most frequently used keywords were "Artificial Intelligence" (247 incidents), "Plagiarism" (111 incidents) and "Research Ethics" (58 incidents).

Keyword clusters indicate a strong focus on AI-powered tools, academic integrity. And research ethics, including "ChatGPT" (171 activities), "machine learning" (56 activities), and "natural language processing" (31 activities), were worth noting). The analysis provides a comprehensive overview of issues in research, emphasizing the importance of AI, ethics, and academic integrity in research (Fig. 3).

Table 5 Most preferred sources

Source title	Publications	% of Total publications
BMJ Open	33	3.40
Lecture Notes in Computer Science Including Subseries Lecture Notes in Artificial Intelligence and Lecture Notes in Bioinformatics	32	3.29
ACM International Conference Proceeding Series	17	1.75
Ceur Workshop Proceedings	13	1.34
Nature	13	1.34
Education Sciences	11	1.13
American Journal of Bioethics	10	1.03
Inorganic and Nano Metal Chemistry	9	0.93
Springer International Handbooks of Education	8	0.82
Communications in Computer and Information Science	7	0.72
Total	153	15.74

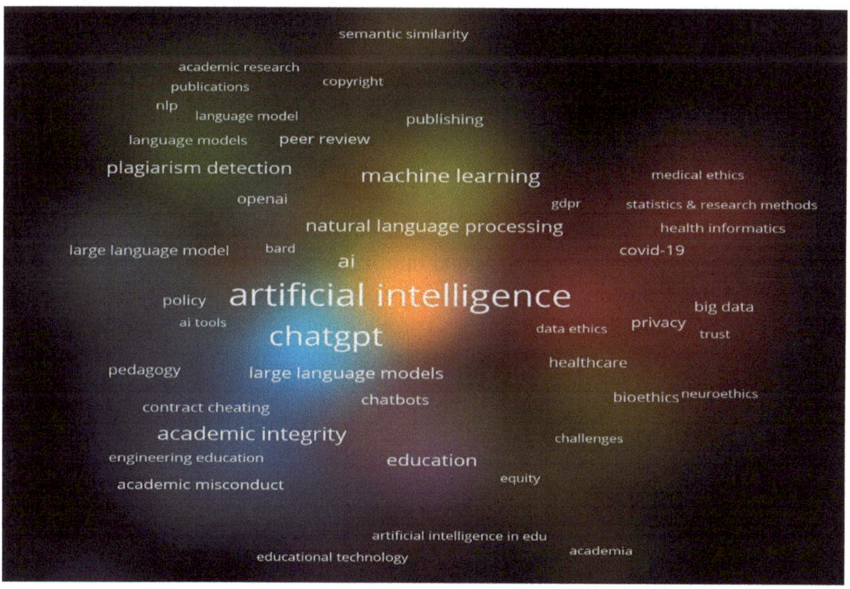

Fig. 3 Keywords co-occurrence

5 Results and Conclusion

The present study is a bibliometric study that provides a comprehensive overview of the influence of AI in the field of research ethics. The study identified a total of 972 such publications, published during a time span of the 44-years (1980–2024) received a total of 13,001 citations, with an average of 13.38 citations per paper. The study revealed that a significant number of publications had been published in recent years, especially in 2023 (369 publications receiving 6929 citations). This indicates increased interest and concern about the ethical implications of AI in research.

These publications were published under a wide variety of document types. Article was the most prominent document type (39.51%), followed by conference papers (22.02%) and editorials (8.95%), reviews and commentaries. Even though Review papers are fewer in number but they have received the highest average citations per publication. The analysis also identifies the most published authors, institutions and countries in this area. It provides insights into research landscapes and potential areas for collaboration.

The United States, the UK, and India were the most prominent countries in the global landscape of AI research and research ethics. It was also revealed that a high degree of international cooperation among the top countries exist, with eight country groups linked by 297 links. 50 countries have at least five papers, demonstrating the strong global interest in research into this field.

The publications were published in different forms of document types, including journals, conference reports, and book collections. Journals such as Nature and the American Journal of Bioethics indicate the importance of analyzing interdisciplinary research. A total of 1,861 author keywords were used in the literature, with 99 keywords used at least 4 times, belonging to 8 different clusters. The most frequently used keywords were "Artificial Intelligence," "Plagiarism," and "Research Ethics." Keyword clusters indicate a strong focus on AI-powered tools, academic integrity, and research ethics. The analysis provides an overview of issues in research, emphasizing the importance of AI, ethics, and academic integrity.

The findings from this study may be helpful for researchers, policymakers and educators. The growing interest in AI and research ethics emphasizes the need to have a strong ethical framework and guidelines in place to address potential bias. The study emphasizes the importance of international cooperation and knowledge dissemination to advance research in this area.

References

Aguilera-Cora, E., Lopezosa, C., Fernández-Cavia, J., & Codina, L. (2024). Accelerating research processes with Scopus AI: A place branding case study.*Revista Panamericana de Comunicación. 6*. https://doi.org/10.21555/rpc.v6i1.3088

Ali, Z., Kefalas, P., Muhammad, K., Ali, B., & Imran, M. (2020). Deep learning in citation recommendation models survey. *Expert Systems with Applications, 162*, 113790. https://doi.org/10.1016/j.eswa.2020.113790

Ampatzoglou, A., Bibi, S., Avgeriou, P., Verbeek, M., & Chatzigeorgiou, A. (2019). Identifying, categorizing and mitigating threats to validity in software engineering secondary studies. *Information and Software Technology, 106*, 201–230. https://doi.org/10.1016/j.infsof.2018.10.006

Caliskan, A., Bryson, J. J., & Narayanan, A. (2017). Semantics derived automatically from language corpora contain human-like biases. *Science, 356*(6334), 183–186. https://doi.org/10.1126/science.aal4230

Doshi-Velez, F., & Kim, B. (2017, February 28). *Towards a rigorous science of interpretable machine learning*. arXiv.org. https://arxiv.org/abs/1702.08608

Elisenda, A. C., Lopezosa, C., & Codina, L. (2024, January 9). *Scopus AI beta: Functional analysis and cases*. http://hdl.handle.net/10230/58658

Jordan, M. I., & Mitchell, T. M. (2015). Machine learning: Trends, perspectives, and prospects. *Science, 349*(6245), 255–260. https://doi.org/10.1126/science.aaa8415

Mehrabi, N., Morstatter, F., Saxena, N., Lerman, K., & Galstyan, A. (2019, August 23). *A survey on bias and fairness in machine learning*. arXiv.org. https://arxiv.org/abs/1908.09635

Strategic Sustainability: Harnessing Alternate Dispute Resolution for Business Conflict Management

Aasritha Vegesena and Pallavi Bajpai

Abstract In this current sustainability-oriented industry, the efficient and responsible management of disputes is an important issue for the competitive advantage of a firm in the long-run. This article focuses on the inclusion of Alternative Dispute Resolution (ADR) mechanisms such as mediation, arbitration, and negotiation in the sustainable business management practices. It argues that by using case studies and industry analysis, alternative dispute resolution can be used to resolve conflicts over environmental, supply chain and stakeholder management aspects while at the same time being consistent with the ESG (Environmental, Social and Governance) standards. The article adopts a qualitative methodology, focusing on economics of mediation strategies such as reducing costs of litigation, increasing equity participation in business through ODR, and protecting the environment. On the societal level, alternative dispute resolution also helps to build community confidence by providing the culturally acceptable, community-inclusive and collaborative approach to problem solving. In summary, this piece of research brings forth a new perspective on the expansive role of alternative dispute resolution on economic advancement towards sustainable development and encourages firms to have alternative dispute resolution principles embedded in the governance systems such that there is equilibrium between profit, people and the planet.

Keywords Alternative dispute resolution (ADR) · Sustainable business management · Environmental · Social · Governance (ESG) · Online dispute resolution (ODR)

A. Vegesena · P. Bajpai (✉)
Bennett University, Greater Noida, India
e-mail: Pallavibajpai@bennett.edu.in

A. Vegesena
e-mail: aasritha.vegesena2@gmail.com

R. S. Mor et al. (eds.), *Sustainable Business Management, Innovation and Technology*, Sustainable Economy and Ecotechnology, https://doi.org/10.1007/978-3-031-98141-8_3

1 Introduction

Conflict is a natural part of social life arising from interaction and this indelible part of human conduct makes resource allocation towards conflict management no longer an option. With the established need for resource allocation towards conflict management due to the inevitable nature of conflict and its impact on overall productivity, alternative dispute resolution provides mechanisms for conflict resolution in a manner cohesive with the organization's interests and other parties involved, facilitating efficiency better than redressal through traditional courts.

Alternate dispute resolution provides a sustainable way to corporate conflict management. This study aims to redefine the boundaries of sustainable corporate practices by identifying how practices of alternate dispute resolution when adopted translate into E.S.G. guidelines (Barrett & Barrett, 2004) which further facilitates stakeholder engagement by directly translating into corporate financial performance.

2 The Nature and Philosophy of Conflict in Corporate Environments

Conflicts are often constituted by the parties involved, to run hand in hand with inconsistency. Etymologically the word conflict is derived from the Latin term "conflictus" which directly translates to "a striking together" or a "collision". Philosophically, Hegelian Dialectics (Barrett & Barrett, 2004) views conflict as a process of thesis, antithesis, and synthesis and the same has been assimilated into modern and broader cultural significances which is no different from the perspective of conflict arising in an organizational set up is viewed with. However, this perspective acts as a roadblock in terms of conflict management (Callan & Thomas, 2009) for efficiency in terms of functionality due to the organizational movement and interaction that characterize conflicts and the adversarial undertones begin to define the relationship between the parties into a conflict.

3 The Impact of Legal Proceedings on Corporate Operations and Stakeholder Engagement

One of the conventional methods for conflict resolution has been to avail traditional courts for redressal, however the flow of interaction amongst the components of a court proceeding irrespective of magnitude is multiplicative in its effect on the adversity (Act, 2013) amongst the parties to the hearing due to the very essence of judicial proceedings that involves legal determination of decisional elements favoring either of the party, and the same is achieved through proceedings that are subject to delays, pendency, backlogs and arrears beyond an individual's control. According to the

World Bank's Ease of Doing Business Report (Dreu & Gelfand, 2008), 'Enforcement of Contracts' metric which ranks countries (Ease of Doing Business Index, xxxx) on the basis of the strength of the regulative environment of the respective country in the year 2020 India ranked 63rd out of 180 countries and with a judge to population ratio of 21 judges per million civilians (Edwards, 1986), the judiciary's efficiency is running at an overridden capacity and this dearth of sufficient judicial force is coupled with statistics that state a duration of 626 days at an average (Edwards, 1986) for corporate conflict resolution and the volatility of this process is further increased by the average time spent by a judge on each listed case in High courts that can be as little as 2–15 min with a case load of 150 cases per day for an average period of 5–5.5 h of work time per judge with intervals of an average gap of 16 days between hearings in the high court of Calcutta and 80 days in the High court of Delhi.

Conventional means of conflict resolution through courts (Environment (Protection) Act, 1986) in Indias are exhaustive in terms of the funding that needs to be pumped into the proceedings that involve an array of expenditures on the lines of court fee, legal fee, witness fee and any other expenses incurred in connection with the proceedings; apart from these expenses the court may order the unsuccessful party to pay the costs of the proceedings to the successful party, this however this is circumstantial and not definite.

This process is not only uneconomical but also contains substantial risk since the possibility of legal proceedings being abortive of the company's stance in a case is of a magnitude that is not dismissive.

Unguaranteed of adversity ingrained in the structural complexity of a conflict (FGV São Paulo Law School et al., 2024) in a corporate setting originating from either within the company or from its interaction with external parties that is further amplified by court proceedings constitutes a wide array of consequences (Friede et al., 2015) stemming from supply chain disruptions, reputational damage, structural disintegration, high employee turnover to name a few that depend on the nature of the dissonance arising in a business coupled with elements of high expenditure and unguaranteed rates of success combined with exhaustive time requirements despite being fruitful cause a blow to the public image of a business, unwanted disclosure of sensitive issue and indicates ineffective dispute management (Greenwood, 2007) on the part of the business due to which the management can be negatively affected leading to decreased stakeholder engagement and support. A shift in organizational policy to risk—averse behavior leading to avoidance of innovation due to fear of potential liability caused by ongoing legal challenges is also a possibility that increases with involvement in litigation (Hegel, n.d.).

An adjunctive tool alterable to fit the needs of the parties and their grievances involved that acts as a mitigator presents itself as alternate dispute resolution, providing businesses with a solution catering to all the shortcomings of conflict resolution through traditional court proceedings.

Alternative dispute resolution is an alternative that views conflicts through a paradigm entirely different from court proceedings. Mechanisms such as mediation, arbitration, negotiation and conciliation with room for hybrid processes that constitute alternative dispute resolution, view and proceed with dissonances in the corporate as disputes and not conflicts. This difference is what makes alternative dispute resolution a lucrative option to an organization due to the very nature of a dispute which is etymologically derived from the Latin term '*Disputare*' which combines the prefix 'Dis' meaning apart or in different directions and the verb '*Putare*' meaning 'to reckon' and therefore literally translating to weigh or calculate arguments in different directions, this cohesive view of inconsistency or disagreements by eliminating the perspective of an opposing and recognizing the difference creating room for negotiation or reaching a middle ground catering to the demands of the disputed parties is an element of "dispute" that has been adopted into modern English usage as well with philosophical roots in ancient debate practices such as the Socratic method involving dialogue instilling disputes to reach deeper truths and Scholasticism, a medieval tradition that used disputation to reconcile theology and philosophy, hinting at an intrinsic structure of communal dialogue with room for negotiation facilitating middle ground achievement through voluntary participation.

4 Exploring Alternative Dispute Resolution Methods: Mediation, Arbitration, Negotiation, and Conciliation

Alternate dispute resolution comes into the picture as a third party-based redressal forum, voluntarily chosen by the disputants to have their views heard and undertake a process that satisfies every claimant involved in a way court proceedings cannot. The following methods are involved (Herremans et al., 2016).

4.1 Mediation

A voluntary process involving a third party who indulges in exploring the differences to find a resolution. The mediator has no authority to impose a settlement, and the process is successful when the disputants mutually find workable solutions. Often used when the disputants have an ongoing relationship they want to protect or preserve (Martinez-Alier, 2021).

Facilitation of conversation and party-driven outcome due to absence of an imposition in terms of settlement are core properties of mediation. The mediator involved must be skilled in terms of counselling, conflict analysis, ethics and consensus building to be able to create trust and confidence between the parties to facilitate effective communication (Porter & Derry, 2012).

4.2 Arbitration

A private process of adjudication in which the claimants choose the decision makers and the rules of procedure, evidence and decision by which their disputes will be resolved. The neutral 3rd party involved in this case is known as an arbitrator who is responsible for going through the evidence presented and makes a binding decision, unlike mediation. The arbitrators involved are selected by the disputants and are often experts in the subject matter of the dispute.

The arbitrator's decision is typically binding and enforceable. Therefore, it is generally final with limited judicial review. With long history of roots in ancient Greece, Rome and Talmud (Rahim, 2023) it was a common practice, particularly among merchants who valued speed and finality. Arbitration gained prominence in the United States of America through labor management relationships initially. However, the first modern laws on arbitration were passed in the 1920s to make arbitration legally valid and enforceable. The arbitrator's binding resolution, speed, efficiency and the disputant's choice in terms of selecting the decision maker and modality of proceedings are some of the key elements of this mechanism that prove to be fruitful for businesses choosing arbitration as a method of dispute resolution (Rosenberg & Folberg, 1994).

4.3 Negotiation

Negotiation is a fundamental form of alternative dispute resolution that involves direct communication between the disputants to reach a voluntary agreement or an attempt to arrange a new combination of their common and conflicting interests. Negotiation is the most basic form of alternative dispute resolution, and it is the precursor to other methods such as mediation and arbitration. Voluntary agreement to hold direct communication through flexible methods to focus on individual interests which will result in a cohesive decision through an informal process with few described rules or procedures characterize negotiation. Competition in negotiatory proceedings is not entirely absent, however, parties are not necessarily adversaries, but rather counterparts working together. Direct control over the process and outcome which is flexible and can be adapted to suit the specific needs of the disputants can help maintain or improve relationships between the parties and at the same time is less expensive than other forms of dispute resolution, such as litigation or arbitration.

4.4 Conciliation

Conciliation is a voluntary process that involves a neutral 3rd party known as the conciliator, who helps disputing parties to communicate, identify issues and explore solutions to resolve their conflict. However, the conciliators' role is to facilitate communication and understanding, however, does not involve imposition of a particular decision or a settlement. Unlike arbitration or litigation, conciliation is not an adjudicative process, but instead helps parties reach their own mutually agreeable solution through a very informal process. While both conciliation and mediation involve a neutral 3rd party, mediators are generally more actively involved in guiding the parties towards settlement, while conciliators are focused primarily on mutually beneficial options. Conciliation satisfies the need for a situation where parties are willing to work together but need a neutral facilitator to guide discussions and help explore options for resolution.

4.5 Integrating Sustainability and Corporate Conflict Resolution: Enhancing Long-Term Business Viability

Sustainability as a concept entails meeting the needs of the present generation without compromising the ability of future generations to meet their own needs. Long term perspectives, interconnectedness, balance and the perspective moving beyond environmentalism are elements that define sustainability in a corporate setting. The concept of sustainability emerged as a way that expanded beyond the scope of environmentalism to include social issues and global economic viability as well thanks to the 1987 Brundtland report (Lok Sabha Secretariat, n.d), this shift could be seen materializing in the form of the "greening label" evolving into "sustainability" which is further translated into business through concepts such as "triple bottom line" incorporating environmental and social consideration along with economic ones (Securities & Exchange Board of India, 2021). Stakeholder engagement is a concept that materialized due to the demand for sustainability by stakeholders.

Social inequalities, climate change and economic disparities defining the twenty-first century can be traced as the reasoning behind why sustainability in corporate practices is no longer a choice and why compliance-based mandates have manifested in the form of multiple statutory requirements such as Companies Act 2013 (Securities & Exchange Board of India, 2015) mandating companies meet specific financial thresholds to allocate at least 2% of their average net profits to CSR activities and Environment Protection Act 1986 (India Cuts Taken & for Disposal of Commercial Cases by50%, 2023) setting standards for pollution control, Energy Conservation Act 2001 mandating energy audits, and SEBI Regulations 2015 mandating top 1000 listed companies to submit Business Responsibility and sustainability reports starting from the financial year 2022–23 (Uwazie, 2011).

Apart from statutory mandates corporate sustainability is also pushed forth through incentivization in the form of tax benefits, subsidies, green bond, sustainability link loans, sustainability awards conferred by Confederation of Indian Industry [CII] facilitating in the enhancement of brand reputation that results in consumer and investor attraction facilitating stakeholder engagement.

Long term business viability, market competitiveness, investor attraction and regulatory compliance are four factors that businesses must not be taken for granted (Uwazie, 2011) so as to grow and due to the persisting state of ecological depletion, social inequality and the interaction with which are controlled and monitored through statutory mandates, incentives, and regulatory mechanisms as mentioned earlier, incorporation of alternate dispute resolution by companies results in contribution to the four factors stated above.

4.6 *The Role of Alternative Dispute Resolution in Promoting Corporate Sustainability Through Environmental Standards*

Alternative dispute resolution when adopted due to its economic efficiency, timely redressal, reduced scope of uncertainties, relationship preservation and ingrained flexibility and adaptability stemming from disputant empowerment throughout the proceedings makes it the call of the hour.

Alternative dispute resolution can contribute to corporate sustainability by an organization or a business due to its overreaching effect on stakeholders, involved disputants and efficiency in addressing environmental and natural resource disputes through Environmental, Social, and Governance standards that are integral to a business strategy in today's time.

Environmental, Social, and Governance standards referred to environmental, social and governance criteria that are used to evaluate a company's performance, and impact in these three areas in terms of sustainable and responsible growth. Investors are increasingly beginning to look at metrics beyond finance to assess a company's sustainability and ethical practice.

4.7 *Alternative Dispute Resolution in Environmental Dispute Resolution: Strengthening Environmental, Social and Governance Compliance and Stakeholder Engagement*

Environmental factors involved in this standard of sustainable performance measurement of a business, or an organization relate to a company's impact on its ecology. It involves subtopics on the lines of carbon emissions, pollution, energy efficiency, water scarcity, deforestation, use of genetically modified organisms and environmental performance etcetera. A company's interaction with the environment is not limited and involves multifaceted dealings with various parties such as ecological beneficiaries, indigenous populations, endangered flora and fauna, cultural sensitivities, activists, raw material supply chains etcetera that open doors to ecological distribution conflicts or EDCs, which are essentially social conflicts caused by environmental degradation or by unequal distribution of environmental resources. The Environmental Justice Atlas documented up to 3100 Environmental Conflicts worldwide as of April 2020; locally affected communities, states, companies, investors, social or environmental movements, environmental defenders, resource extraction victims are often the sects from which the parties involved in this conflict come from there are studies directly linking environmental conflict to industrial economy with statistics indicating that less than 10% of materials and energy are recycled, industrial economy is constantly expanding through extensive material extraction at commodity frontiers by appropriating either new natural resources through territorial claims and land grabs or exploitation of existing sites. EDCs often arise from disparities in access to natural resources or unequal distribution of burdens of environmental pollution and relate to exercise of power stemming from different social actors when they enter disputes over access to or impacts on natural resources due to the multiplicity of culture sensitive factors involved in environmental distribution conflicts alternate dispute resolution provides an appeal of amicable resolution of recurring environmental claims and litigation. Environmental dispute resolution has been evolved, altered to fit the needs of ecological distribution conflicts through principles of alternate defense resolution in the 1960s in US and Canada due to increasing environmental litigation with equally strong developments in usage of alternate dispute resolution for environmental conflict resolution in Japan and South Korea. The flexibility and context specific approaches needed to ensure a fair and sustainable outcome in environmental disputes is the demand answered by dispute redressal through alternative dispute resolution mechanisms and the same can lead to effective, efficient and favorable conflict resolution with respect to environmental disputes of a business. By preventing negative brand reinforcement and economic exhaustion brought by court trials this assimilation of alternative dispute resolution mechanisms in ecological distribution conflicts contributes positively to stakeholder engagement due to it translating to proactivity with respect to the organization in terms of conflict management. The same, however, has not been reflected in international legal institutions catering to environmental disputes such as ICJ, WTO AND ITLOS since they have had limited success in resolving environmental disputes often due to a lack of specialized expertise and a focus on strict legal norms rather than environmental imperatives. This proactivity is highly likely to be perceived in a good light by stakeholders further facilitating support and catering to the environmental measure of Environmental, Social, and Governance standards.

4.8 *Alternative Dispute Resolution and Social Responsibility: Bridging Disputes for Better Business Relationships*

The cohesive assessment of dissonance in a business setting by taking into consideration all the disagreements to reach a middle ground by businesses is the very crux of alternate dispute resolution, enhancing social responsibility by providing a platform for addressing disputes with intentions of building better relationships moving beyond mere conflict control and thereby delivering effective conflict management. The accessibility of alternative dispute resolution mechanisms provided to the disputants is of immense importance that can be leveraged by those with scarce resources and promotes an extension of human rights through provision of a level playing field as opposed to litigation that is adversarial. The scope that alternative dispute resolution brings with it in terms of the type of parties that can be involved is dependent on the type of dispute which contributes towards the social factor of Environmental, Social, and Governance guidelines as observed in the case Chloro Controls India (P) Ltd. v. Severn Trent Water Purification Inc., (2013) 1 SCC 641 [2012 INSC 436] and Cox and Kings Ltd v. SAP India Pvt Ltd, (2022) 8 SCC 1, para 54 [2022 INSC 523] where the Court looked at the tight group structure and single economic unit when deciding that a non-signatory could be bound by an arbitration agreement while Sukanya Holdings V. Jayesh H Pandya (2003) 5 SCC 531 [2003 INSC 230] and Indowind Energy Ltd v. Wescare (India) Ltd (2010) 5 SCC 306 [2010 INSC 246] are cases which limited parties to an arbitration agreement to its signatories placing emphasis on consent as required for proper redressal of the dispute as reemphasized by Oil And Natural Gas Corporation LTD. Versus M/S Discovery Enterprises PVT. LTD. & ANR. SCR Citation: [2022] 4 S.C.R. 926.

The Administrative Conference of the United States 1968 was established by the Congress to study efficiency, adequacy and fairness of administrative agencies in the 1980s 'ACUS' became the strongest supporter of alternative dispute resolution by 1990 it had issued 17 reports supportive of alternative dispute resolution in administration. Although this study focused on assimilation of alternative dispute resolution mechanisms into federal structures and governance, the administrative efficiency derived through the same can be replicated in administrative units of businesses as well as company leadership structure and internal controls, negatively affected due to external factors such as negative brand reinforcement distrust and prolonged litigation.

Alternative dispute resolution can make the rationale behind decisions more accessible and understandable; this is particularly relevant in environmental disputes where public access to information is crucial.

Mechanisms like arbitration, mediation and conciliation etcetera, increase accountability due to dialogue exchange through direct communication despite the intervention of neutral 3rd parties, demands justification of positions and impacts of their actions on others. The structural mechanism of alternate dispute resolution facilitates parties to take responsibility for their behavior which is highly effective when it involves a neutral 3rd party.

4.9 Stakeholder Engagement Through Alternative Dispute Resolution: Enhancing Organizational Governance and Environmental, Social, and Governance Compliance

Incorporation of alternate conflict resolution into conflict resolution by businesses and organization contributes to sustainability measurable through Environmental, Social, and Governance guidelines, materializing into stakeholder engagement, which is a core aspect of alternate dispute resolution since involving stakeholders in the resolution process through different participatory tools to share representation of various interest groups and perspectives forms the very basis of this mechanism.

Stakeholder engagement is the stakeholder involvement in a positive manner into the organizational functions, which is a crucial aspect displaying how a company interacts with and addresses the needs of various interest groups and stakeholders who are individuals and groups that can be affected or are affected by the company's action. Key components of stakeholder management involve stakeholder identification and understanding the stakeholders needs, communication and building relationships through balancing conflicting claims, which evidently form a reiteration of the essentials of alternate dispute resolution and therefore facilitate the organization in better governance through contributing towards stakeholder engagement, fulfilling the governance aspect of Environmental, Social, and Governance guidelines.

Sustainability reports are used as a tool by companies for stakeholder engagement. The characteristics of sustainable reporting include directness of communication, clarity and stakeholder identification, deliberateness of collecting feedback, broadness of stakeholder inclusiveness, utilization of stakeholder engagement for learning. All the above set factors are also inclusive alternative dispute resolution principles further reinforcing the favorable magnitude of contribution alternative dispute resolution poses towards sustainable governance.

4.10 Alternative Dispute Resolution and Corporate Financial Performance: The Positive Link to Environmental, Social, and Governance Compliance

Corporate financial performance is a measure of a company's financial health over a period used to determine how well it generates revenue and profit, assessed by looking at the company's financial statements.

Through statistics provided by a metanalysis of over 2000 empirical studies on the relationship between Environmental, Social, and Governance criteria and corporate financial performance a study provided a comprehensive overview of academic research on the topic and determined a positive ESCP relationship by approximately 90% of the studies finding a non-negative relation with majority reporting positive findings. Vote count studies which assessed the direction of the relationship found that above 48% of the study showed a positive relationship, while 10% showed a − 1.

The met analysis also statistically combined the results of multiple studies indicating a larger share of about 95.8% of the studies indicating a positive correlation.

While wider components of Environmental, Social, and Governance criteria all showed a positive relationship, environmental and governance factors showed a slightly more positive relationship than the social factors. Regionality also influenced positivity in terms of emerging markets, showing a higher share compared to developed markets. However, positive Environmental, Social, and Governance, corporate financial performance relationship appears to be stable since the mid-1990s with no consistent learning effects detected. The possibility of investors aligning their interests with the border goals of society by integrating Environmental, Social, and Governance criteria into their investment processes is of beneficiary magnitude for corporates to consider Environmental, Social, and Governance compliance.

There has been a distinction between portfolio studies and non-portfolio studies in the context of Environmental, Social, and Governance and corporate financial performance relationship, highlighting differing methodologies and results. Non portfolio studies tend to report a more positive relationship between Environmental, Social, and Governance performances and corporate financial performance generation since these studies may focus on operational efficiencies, risk management improvements or long term value creation driven by Environmental, Social, and Governance initiatives, often bypassing the practicalities and constraints of real world investing such as transaction costs or investor behavior while portfolio studies focus on real world investment products like Environmental, Social, and Governance mutual funds or portfolios built around Environmental, Social, and Governance criteria, these studies tend to exhibit weaker, neutral or mixed relationship between Environmental, Social, and Governance and corporate financial performance, possible reasons could be the Environmental, Social, and Governance funds incurring higher costs related to research, screening and portfolio rebalancing while overvaluation of Environmental, Social, and Governance assets due to high demand and challenges in aligning Environmental, Social, and Governance ratings with financial outperform and may reflect short term volatility overshadowing long term Environmental, Social, and Governance benefits. Therefore, revealing practical challenges and nuances in translating Environmental, Social, and Governance criteria compliance-based advantages into immediate financial results in investment context.

5 Understanding Sustainability Reporting System in India

Sustainability in India is no longer a mere proactive approach to poaching green investors and this statement is substantiated by identifying two trends in corporate sustainability reporting ecosystem in India i.e., the Business Responsibility Sustainability Report timeline and the recent developments in respect of sustainability report mandation in G20 Sustainable Finance Working Group (SFWG) July 2024.

5.1 Business Responsibility Sustainability Report

At a nascent stage National Voluntary Guidelines on Social Environmental and Economic Responsibilities of Business (NVGs) governed sustainability reporting in 2011, however were later shifted into a mandate in 2012 for the top100 listed companies in the form of Business Responsibility Report which later extended its scope to top 500 listed companies followed by top 1000 listed companies in 2019. This shift from voluntary adherence to a statutory mandate can be seen accelerating at a pace that companies need to start catching up with.

5.2 Sustainability Reporting in the Digital Age—G20 Sustainable Finance Working Group

Sustainability reporting in the Digital age is a report submitted to the G20 Sustainable Finance Working Group (SFWG) July 2024. An analysis of which is enough to indicate how the environment reporting system in India is going to transform soon.

The report essentially talks about the following measures to be initiated as follows:

1. Developing proportionate sustainability reporting standards for small and medium sized enterprises along with emerging markets and developing economies.
2. A globally standardized sustainability reporting system
3. A shift from voluntary disclosure to mandatory regulation-based frameworks

Out of many that would make sustainability reforms in corporate structures the priority for the management.

Therefore, leveraging Alternate dispute resolution for corporate conflict management eases this burden on the management's shoulders while contributing to both statutory compliance that is going to tighten in the years to come alongside effective and efficient conflict resolution.

6 Conclusion

The transformative approach of business conflict management through alternative dispute resolution, simultaneously catering to corporate sustainability provides the organization with evident benefits of resource efficiency in conflict management through significant contributions to long-term prospective sourced from stakeholder engagement facilitation.

Steps such as the institutionalization of alternative dispute resolution in corporate governance by establishing industry specific alternative dispute resolution forums that bring together corporates, regulators and community representatives to collaboratively address systemic conflict or recurring issues could further contribute to Environmental, Social, and Governance criteria through social inclusion resulting in Environmental, Social, and Governance criteria compliance, democratizing access to conflict management and bridging geographical divides which are essential characteristics of online dispute resolution that can be leveraged by businesses striving to align with Environmental, Social, and Governance goals.

Online dispute resolution will as a revolutionary mechanism for resolving business conflicts, particularly in the context of disputes involving supply chains and cross border transactions, along with environmental concerns. Since resolution timelines and access to justice through a sustainability driven business environment satisfy the demands of speed and accessibility that are highly critical in disputes of such nature, online dispute resolution mechanisms can provide businesses with conflict resolution mechanisms that are further efficient in terms of reducing carbon economic efficiency and increased accessibility.

The consequent development of industry specific online dispute resolution platforms such as logistics, manufacturing or energy industry specific online dispute resolution mechanisms by leveraging domain expertise, facilitates alternative dispute resolution mechanisms translating into Environmental, Social, and Governance standard compliance through AI based tracking. Since this method, if promoted through conducting targeted awareness campaigns, backed by user friendly interfaces and supported through integration with regulatory frameworks, eases the company's efforts in terms of integration amongst conflict management and stakeholder engagement resulting in financial performance through Environmental, Social, and Governance compliance.

Viewing alternate dispute resolution mechanisms as methods of operational efficiency in terms of conflict management would be an understatement in terms of the benefits that could be leveraged such as strengthening a company's commitment to sustainable business practices through alignment with Environmental, Social, and Governance mandates results in alternate dispute resolution serving as a vital pillar to achieve long term corporate resilience and the creation of an equitable corporate ecosystem.

References

Barrett, J. T., & Barrett, J. P. (2004). *A history of alternative dispute resolution: The story of a political, cultural, and social movement*. Jossey-Bass.

Callan, S. J., & Thomas, J. M. (2009). CFP and CSP: An update and reinvestigation. *Corporate Social Responsibility and Environmental Management, 16*(2), 61–78.

Companies Act, 2013 (India). https://www.mca.gov.in/Ministry/pdf/CompaniesAct2013.pdf

De Dreu, C. K. W., & Gelfand, M. J. (2008). *The psychology of conflict and conflict management in organizations*. Lawrence Erlbaum Associates, pp. 3–54

Ease of doing business index. The World Bank. https://data.worldbank.org/indicator/IC.BUS.EASE.XQ. Last visited [insert date]

Edwards, H. T. (1986). Alternative dispute resolution: Panacea or anathema? *Harvard Law Review, 99*, 668.

Environment (Protection) Act, (1986). (India). https://www.indiacode.nic.in/bitstream/123456789/4316/1/ep_act_1986.pdf

FGV São Paulo Law School et al. (2024) *Sustainability reporting in the digital age: Overcoming challenges for SMEs and EMDEs*. G20 Sustainable Finance Working Group Report.

Friede, G., et al. (2015). ESG and financial performance: Aggregated evidence from more than 2000 empirical studies. *Journal of Sustainable Finance & Investment, 5*(4), 210–233.

Greenwood, M. (2007). Stakeholder engagement: Beyond the myth of corporate responsibility. *Journal of Business Ethics., 74*(4), 315–327.

Hegel, G. W. F. (n.d.). *Hegel's Dialectics* (19th century) (unpublished manuscript or original philosophical concept, no formal citation format in Bluebook; Consider citing a critical edition if available)

Herremans, I. M., et al. (2016). Stakeholder relationships, engagement, and sustainability reporting. *Journal of Business Ethics, 138*(3), 417–435.

Martinez-Alier, J. (2021). Mapping ecological distribution conflicts: The EJ Atlas. *The Extractive Industries and Society., 8*(4), 100883.

Porter, T., & Derry, R. (2012). Sustainability and business in a complex world. *Business and Society Review, 117*(1), 33–53.

Rahim, M. A. (2023). *Managing conflict in organizations*. Routledge.

Rosenberg, J. D., & Folberg, H. J. (1994). Alternative dispute resolution: An empirical analysis. *Seattle UL Review, 46*, 1487.

Lok Sabha Secretariat. *Unstarred question no. 1335: Commercial dispute resolution*, https://sansad.in/getFile/loksabhaquestions/annex/1715/AU1335.pdf?source=pqals. Last visited [insert date]

Securities and Exchange Board of India, (2021) *Business responsibility and sustainability reporting by listed entities*. https://www.sebi.gov.in/sebi_data/commondocs/may-2021/Business%20responsibility%20and%20sustainability%20reporting%20by%20listed%20entitiesAnnexure2_p.PDF

Securities and Exchange Board of India, (2015). *Listing obligations and disclosure requirements regulations, 2015* (Amended Sept. 20, 2023). https://www.sebi.gov.in/legal/regulations/sep-2023/securities-and-exchange-board-of-india-listing-obligations-and-disclosure-requirements-regulations-2015-last-amended-on-september-20-2023-_77239.html

India Cuts Time Taken for Disposal of Commercial Cases by 50%, (2023). Times of India (Mar. 15, 2023). https://timesofindia.indiatimes.com/india/india-cuts-time-taken-for-disposal-of-commercial-cases-by-50/articleshow/93696747.cms

Uwazie, E. E. (2011) Alternative dispute resolution in Sub-Saharan Africa. In Moffitt, M. L., & Bordone, R. C. (Eds.), *The handbook of dispute resolution* p. 517.

Uwazie, E. E. (2011). Alternative dispute resolution in Sub-Saharan Africa. In Moffitt, M. L., & Bordone, R. C. (Eds.), *The handbook of dispute resolution*, p. 517.

Startup Ecosystem, Innovation, and Entrepreneurship in India

Sanjay Kumar Mangla and Yatika Gupta

Abstract There is a global focus on promoting entrepreneurship by almost all countries as it not only helps governments to tackle unemployment problems but contributes significantly to the nations' economic size. Therefore, governments are taking a bundle of efforts to boost entrepreneurship including promoting innovation and startup cultures in their respective countries. India is also a front runner in these aspects and has introduced many such initiatives including Startup India and increasing expenditure on research & development and producing more patents. This paper is an empirical investigation of analysing impacts of startup ecosystem and innovation on entrepreneurship in the country.

Keywords Start-up ecosystem · Patents · R&D expenditure · Innovation · Entrepreneurship · India

1 Introduction

The global economic landscape is changing rapidly, thanks to technological advancements, societal changes and economic shifts. Besides, land, labour and capital, knowledge has also become primary driver of economic growth. Acquisition, dissemination and application of knowledge aids in generating new products or services or improving the existing ones. A knowledgeable workforce acts as the foundation for innovation.

Innovation serves as a vehicle that upgrades idea and acts as a channel for realization of knowledge spillovers. The term 'Innovation' may refer to include process of converting an idea into product or service, a new way of presenting a product or

S. K. Mangla (✉)
Symbiosis Institute of Management Studies, Symbiosis International (Deemed University), Pune, India
e-mail: sam211285@gmail.com

Y. Gupta
Faculty of Management, Symbiosis International (Deemed University), Pune, India

R. S. Mor et al. (eds.), *Sustainable Business Management, Innovation and Technology*, Sustainable Economy and Ecotechnology, https://doi.org/10.1007/978-3-031-98141-8_4

service (designing, marketing), construction of a new industrial equipment. Business' ability to launch new products faster than the competitors is a disguised and most distinctive competitive advantage. Research has shown that companies that innovate continuously earn double than their competitors who innovate lesser.

Innovation drives entrepreneurship, creating a dynamic nexus that drives economic growth. Entrepreneurs are the ones who identifies and exploit opportunities. As Schumpeter describes, an entrepreneur creates new combinations- new products, new markets, new materials. Innovation provides creative ideas, better products, or improved processes that entrepreneurs bring to the market. Carayannis et al. (2015) asserted that entrepreneurs, in turn, renders these innovations into viable businesses that can upend traditional markets disrupt existing industries, increase market penetration, generate employment and spur economic growth. At the same time, it is also true that entrepreneurs are responsible for innovations. Innovation requires research and development (R&D) that helps in the process of experimentation and market engagement. While all other have diminishing returns, investments in knowledge have increasing returns. Entrepreneurs or start-ups[1] develop new insights, refine business models, and drive continuous improvement. This iterative process of learning and adaptation is necessary for the long-term success of businesses and the industries they operate within. This makes entrepreneurship a vital ingredient for economic growth for both developed and developing economies.

Governments play significant role in enabling and promoting the start-up ecosystem. The administration across nations provides digital infrastructure, funding at lower interest rates, regulatory support and tax holidays to encourage innovation and entrepreneurship. The start-up environment in India is teeming with innovation and robust spirit of entrepreneurship, making India as the third largest startup ecosystem globally.[2] The Government of India introduced Start-up India initiative in 2015 that aims to create a robust environment to support innovation and startups that propel long-term economic growth and create significant number of job opportunities. Substantial progress has been made since then. As per the latest data released by Department of Promotion of Industry and Internal Trade (DPIIT), there exists around 1.45 lakh recognised start-ups, employing more than 12 lakh individuals.

Simplifying and digitizing the procedure for formation of a business has been one of the cornerstones of the initiative encouraging start-ups. Since inception, 53 regulatory reforms have been enacted to boost the startup ecosystem in India. Through SPICe (Simplified Proforma for Incorporating Company electronically) platform, start-ups can easily incorporate their business on a single web form and there are no incorporation fees for businesses have capital up to Rs. 15 lakhs. Tax deductions on profits for 3 consecutive years under certain conditions, 80% rebate on patent filing

---

[1] When a business is in early stages of development, it is termed as start-up. It is a small business that often focuses on brining a unique product of service to the market to later establish itself as a successful company.

[2] https://www.startupindia.gov.in/content/sih/en/international/go-to-market-guide/indian-startup-ecosystem.html.

and a 50% rebate on trademark filing compared to other businesses are a few fiscal provisions provided by under the scheme.[3]

Further, the scheme offers financial and infrastructural assistance as well. The Initiative for Developing and Harnessing Innovations (NIDHI) has developed close to 200 technology business incubators that provide fundings to innovators. As of December 31, 2023, funding of Rs. 841.8 crores have already been approved under the Scheme. Financial and technological support for startups working on cutting-edge technologies like the Internet of Things, Artificial Intelligence (AI), Blockchain, Robotics, are provided under the enhanced Technology Incubation and Development of Entrepreneurs (TIDE 2.0) scheme. Furthermore, the government has initiated reforms that propel foreign partnerships and provide global exposure to the new entrepreneurs. Relaxation of FDI policy to allow 100% foreign direct investment (FDI) in several sectors. Additionally, start-ups are exempted from angel tax regime since its inception. There are no taxes and duties on exports under the Remission of Duties and Taxes on Exported Products (RoDTEP) scheme for start-ups. These reforms have paved way for Indian start-up ecosystem in a move to take the Indian economy to new heights in the field of innovation.

Thus, against this background of increasing government initiatives to boost startup ecosystem and innovation culture in India, it is imperative to examine the impact of these two critical factors on entrepreneurship which is important for generating employment in the country.

2 Related Literature

The start-up ecosystem in India is largely driven by the vibrant connection between innovation and entrepreneurship. Extensive work has been done in regard to ascertaining the relationship between the variables and other mediating factors affecting them. Veeraraghavan (2009) emphasized the role of factors including family background and risk-taking in shaping entrepreneurial capabilities. González-Pernía et al. (2015) observed a limited connection between knowledge innovation and entrepreneurship in context of developing countries. Juliana et al. (2021) found a strong relationship between creativity and innovation with significant role played by technological advancement. However, this did not translate into the development of entrepreneurship in Nigeria. The authors suggested government support in this direction to achieve economic success.

On the contrary, researchers have found significant impact of knowledge and creativity on entrepreneurship. Hussain et al. (2011) used correlation and regression analysis to find the impact of innovation, technology and economic growth on the entrepreneurial activities. The results showed that these variables are highly correlated and significantly impact entrepreneurship. Akinwale et al. (2020) used regression analysis to find that significant and positive impact of innovation on

[3] https://www.investindia.gov.in/team-india-blogs/role-government-initiatives-boosting-startups.

entrepreneurship. However, the study also found that innovation does not significantly impact economic growth as against entrepreneurship. Farinha et al. (2018) used descriptive statistics, structural equation modelling (SEM), and hierarchical cluster analysis to investigate the relationship between innovation, entrepreneurship, and economic growth across countries at various development levels. It was found that innovation is necessary for economic competitiveness. Through literature review, Botelho et al. (2021) underscored that economic growth is driven by high-growth startups rather than the overall quantity of new firm entries.

Research have also been carried out on determining the influence of certain factors affecting entrepreneurship locally and internationally. Jabłońska and Stawska (2020) found that R&D expenditure and remuneration in the Polish economy are important factors in influencing the development of entrepreneurship domestically which is then positively transferred to the growth of international entrepreneurship. Amponsah and Ahmed (2017) used principal component factor analysis to understand the factors affecting success of entrepreneurship in Dubai. The findings indicated that visionary leadership and government support are critical for the success of entrepreneurship in emerging markets.

Dahlstarnd and Stevenson (2010) emphasized the need to integrate innovation and entrepreneurship policies, highlighting the existing lack of coordination between these areas. They advocated the concept of innovative entrepreneurship policy that underlines the role of two in economic growth and development. Ruggieri et al. (2018) suggested that adoption of digital platforms can substantially benefit emerging start-ups to achieve scalability and unlock new revenue streams. Szirmai et al. (2011) pointed at the need for direct and indirect government support for innovation including providing ease of doing business, subsidies for R&D, infrastructure improvements, technical advancements. Fiorentino et al. (2021) and (Botelho et al., 2021) found that innovation significantly supports growth of new businesses especially start-ups. They found that R&D expenditures have larger impacts on start-up growth than patents. The study further underscored the importance of policy support for faster growth of innovation-based start-ups.

Anokhin and Schulze (2009) highlighted the role of corruption on the two variables. They found that reducing corruption is linked to higher levels of innovation and entrepreneurship, while the absence of trust increases transaction costs and hampers productivity and investment. Mazla et al. (2020) underscored the importance of creativity and innovation in entrepreneurship. The study explored the role of creativity in improving business practices and overall economic development. Kreiterling (2023) studied the impact of digital innovation on entrepreneurship. Findings indicated that government influence significantly impacts entrepreneurship sustainability and innovation boosts productivity, especially in manufacturing domain. Kraus et al. (2021) investigated the contribution of innovation and knowledge to entrepreneurship and regional economic development. The study underscored the importance of public policy in fostering entrepreneurial growth, suggesting that regions should strategically develop conducive environment for innovation to boost growth and support start-ups.

Thus, all the above-mentioned literature lacks in an empirical investigation on the effect of innovation and start up ecosystem on entrepreneurship in India.

3 Research Gap and Objectives

The review of literature points towards the significant and positive link between innovation and entrepreneurship, more pronounced for the businesses that are in the early stages of their lifecycle. Additionally, the researchers have underscored the need of substantial government support in terms of provision of appropriate infrastructure and policy reforms including adoption of digital platforms and fundings. While this has been emphasized widely, yet not much work has been carried out in assessing the impact of the same especially, in developing countries like India. The present paper thus, tries to evaluate the startup ecosystem, innovation, and entrepreneurship in India along with examining the effect of economy size, innovation, and startup ecosystem on entrepreneurship in India.

4 Data and Research Framework

This paper relies on the data for the period from 2006 to 2022 (based on availability) obtained from World Bank Database where entrepreneurship, economy size, and innovation are measured as 'new businesses registered'; 'gross domestic product (GDP)'; and expenditure on research and development, and patents (patents applications by residents and patent applications by non-residents), respectively. Startup ecosystem has been measured as a dummy variable for 'Startup India' initiative announced by the Government of India on August 15, 2015, and implemented on January 16, 2016. Further, to investigate the impact of economy size, innovation, and startup ecosystem on entrepreneurship; this paper estimates the following function and equation:

$$\begin{aligned} NBRit = &f(GDPit, R\&DEXPit, PATENTAPPRit,\\ &PATENTAPPNRit, DUMMYSTARTUP - INDIAit) \end{aligned} \tag{1}$$

$$\begin{aligned} NBRit = &b0 + b1GDPit + b2R\&DEXPit\\ &+ b3PATENTAPPRit\\ &+ b4PATENTAPPNRit\\ &+ b5DUMMYSTARTUP - INDIAit + eit \end{aligned} \tag{2}$$

Here *NBR* = number of new businesses registered; *GDP* = gross domestic product, measured in current US$ billion; *R&DEXP* = research and development expenditure, measured in current US$ million; *PATENTAPPR* = patent applications by residents; *PATENTAPPNR* = patent applications by non-residents; *DUMMYSTARTUP*−*INDIA* = dummy variable for Startup India initiative which takes a value of 0 for years before 2016 and a value of 1 for years 2016 and thereafter; *e* = error term; and *i* and *t* are standard subscript representing country (India) and time.

5 Empirical Estimation and Findings

Key summary statistics of variables under study and their correlation matrix are given in Tables 1 and 2, respectively, while their relationships through scatter plots are shown in Fig. 1. The correlation matrix and scatter plots show that entrepreneurship variable i.e. NBR has a significant positive correlation with all its explanatory variables considered under this study, however, it does not establish that these explanatory variables significantly decide NBR, for which Eq. (2) has been estimated using regression.

Table 1 Summary statistics of variables under study

Variable	Obs	Ref. Period	Mean	Std. Dev	Min	Max
New businesses registered, number	17	2006–2022	100202.20	36037.18	49721	173114
GDP, current US$ billion	17	2006–2022	2100.23	715.59	940.26	3353.47
R&D exp, current US$ million	15	2006–2020	13884.83	3223.84	7566.55	18698.55
Patent app, residents	16	2006–2021	12594.69	6110.56	5686	26267
Patent app, non-residents	16	2006–2021	31559.13	3095.96	23242	35306

Source Authors' calculations

Table 2 Correlation among variables under study

	NBR	GDP	R&DEXP	PATENTAPPR	PATENTAPPNR
NBR	1.0000				
GDP	0.9196*	1.0000			
R&DEXP	0.8799*	0.9902*	1.0000		
PATENTAPPR	0.9272*	0.9344*	0.8923*	1.0000	
PATENTAPPNR	0.8002*	0.7783*	0.7916*	0.6713**	1.0000

Source Authors' calculations. * Significant at 1%, ** significant at 5%

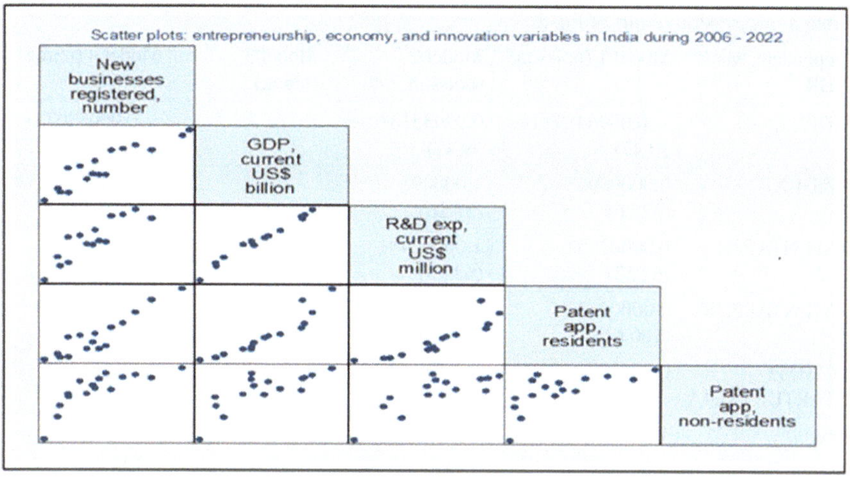

Fig. 1 Relationships among entrepreneurship, economy, and innovation variables in India during 2006–2022. *Source* Authors

For empirically examining the impact of economy, innovation, and startup ecosystem on entrepreneurship in India through estimating Eq. (2), Poisson regression is a suggested method as the dependent variable i.e. *NBR* is a positive count number, however, its variance is much higher than its mean (Table 1), therefore, it suffers from 'over dispersion' and Poisson regression results are not statistically valid. At the same time, *NBR* data does not have any zero value, therefore does not suffer from 'zero-inflation' (there are more zero values than expected). Thus, considering the presence of over dispersion and absence of zero inflation; the statistically suggested method for count data regression is negative binomial regression (nbreg).

The results of Poisson model are presented in Model 1 (with coefficients) and Model 2 (with incidence—rate ratios) whereas results of nbreg are presented in Model 3 and Model 4 (Table 3). The results of nbreg suggest that patent applications (both residents and non-residents) and startup ecosystem have statistically significant positive impact on boosting entrepreneurship activities in India while size of the economy (i.e. GDP) and expenditure on R&D do not have any significant impact on entrepreneurship.

6 Conclusion and Recommendations

The government of India has made commendable efforts to promote innovation and entrepreneurship in the country, especially during the last 10 years. Providing self-employment to its people has been among the priorities of the government and the country has seen some remarkable success also on this front. 'The number of patents (both residents and non-residents) applications has increased from 28,928 in 2006

Table 3 Regression results of Eq. 2

Dependent Var: NBR	Model 1 (poisson)	Model 2 (poisson, irr)	Model 3 (nbreg)	Model 4 (nbreg)
GDP	– 0.0006489* (– 39.47)	0.9993513* (– 39.47)		– 0.0005307 (– 1.64)
R&DEXP	0.0000896* (32.26)	1.00009* (32.26)		0.0000706 (1.33)
PATENTAPPR	0.0000285* (62.53)	1.000029* (62.53)	0.0000192* (2.77)	0.0000265** (2.42)
PATENTAPPNR	0.0000643* (100.61)	1.000064* (100.61)	0.0000576* (6.69)	0.0000618* (4.89)
DUMMY_ STARTUP INDIA	0.226212* (54.24)	1.253841* (54.24)	0.1312303*** (1.80)	0.2255504** (2.52)
Obs	15	15	16	15
Pseudo R^2	0.9414	0.9414	0.1222	0.1207
Deviance goodness-of-fit	6003.366*	6003.366*		
Pearson goodness-of-fit	5988.998*	5988.998*		
chibar2(01) for Likelihood-ratio test of alpha = 0			8041.44*	5897.00*

Source Authors' calculations. * Significant at 1%, ** significant at 5%, *** significant at 10%. Values in parentheses are z-statistics

to 43,031 in 2013 and to 61,573 in 2021. While R&D expenditure during the same period has increased from USD 7.57 billion to USD 13.12 billion and USD 17.29 billion (2020). While new businesses registered, during the same period, increased from 49,721 to 91,841 and to 173,114 (2022)'.[4]

The empirical investigations of this paper suggest that patent applications and startup ecosystem have a significant positive impact on boosting entrepreneurship in the country while R&D expenditure, though, also have a positive association with entrepreneurship, but its coefficient is not statistically significant. This implies that most of the R&D expenditure is either conducted in already established companies or it is not getting converted into registering new firms. The paper suggests that government should continue and expand its initiatives such as Startup India and support to file more patents as these significantly determine entrepreneurship in the country.

[4] This data has been obtained from World Bank database.

References

Akinwale, Y. O., Alaraifi, A. A., Ababtain, A. K. (2020). Entrepreneurship, innovation, and economic growth: Evidence from Saudi Arabia. In: Bilgin, M., Danis, H., Demir, E., Tony-Okeke, U. (Eds.) *Eurasian Economic Perspectives*, vol. 15/1, pp. 45–60. Cham: Springer. https://doi.org/10.1007/978-3-030-48531-3_3

Amponsah, C. T., & Ahmed, G. (2017). Factors affecting entrepreneurship in emerging economies: A case of Dubai. *Journal for International Business and Entrepreneurship Development, 10*(2), 120–137. https://doi.org/10.1504/JIBED.2017.084302

Anokhin, S., & Schulze, W. S. (2009). Entrepreneurship, innovation, and corruption. *Journal of Business Venturing, 24*(5), 465–476. https://doi.org/10.1016/j.jbusvent.2008.06.001

Botelho, T. L., Fehder, D., Hochberg, Y. (2021). Innovation-driven entrepreneurship. *NBER Working Paper Series*, No. 28990. National Bureau of Economic Research. https://doi.org/10.3386/w28990

Carayannis, E. G., Samara, E. T., Bakouros, Y. L. (2015). *Innovation and entrepreneurship: Theory, policy, and practice.* New York: Springer. https://doi.org/10.1007/978-3-319-11242-8

Dahlstrand, A. L., & Stevenson, L. (2010). Innovative entrepreneurship policy: Linking innovation and entrepreneurship in a European context. *Annals of Innovation & Entrepreneurship, 1*(1), 5602. https://doi.org/10.3402/aie.v1i1.5845

Farinha, L., Ferreira, J. J. M., & Nunes, S. (2018). Linking innovation and entrepreneurship to economic growth. *Competitiveness Review, 28*(4), 451–475. https://doi.org/10.1108/CR-07-2016-0045

Fiorentino, R., Longobardi, S., & Scaletti, A. (2021). The early growth of start-ups: Innovation matters. Evidence from Italy. *European Journal of Innovation Management, 24*(5), 1525–1546. https://doi.org/10.1108/EJIM-02-2020-0057

González-Pernía, J. L., Jung, A., & Peña, I. (2015). Innovation-driven entrepreneurship in developing economies. *Entrepreneurship and Regional Development, 27*(9–10), 555–573. https://doi.org/10.1080/08985626.2015.1075602

Hussain, M. F., Afzal, A., Asif, M., Ahmad, N., & Bilal, R. M. (2011). Impact of innovation, technology, and economic growth on entrepreneurship. *American International Journal of Contemporary Research, 1*(1), 45–50.

Jabłońska, M., & Stawska, J. (2020). The key factors affecting entrepreneurship: A comparative analysis. *Zbornik Radova Ekonomskog Fakulteta u Rijeci: Časopis Za Ekonomsku Teoriju i Praksu, 38*(1), 125–146. https://doi.org/10.18045/zbefri.2020.1.125

Juliana, N., Hui, H., Clement, M., Solomon, E., & Elvis, O. (2021). The impact of creativity and innovation on entrepreneurship development: Evidence from Nigeria. *Open Journal of Business and Management, 9*, 1743–1770. https://doi.org/10.4236/ojbm.2021.94095

Kraus, S., McDowell, W., Ribeiro-Soriano, D. E., & Rodríguez-García, M. (2021). The role of innovation and knowledge for entrepreneurship and regional development. *Entrepreneurship and Regional Development, 33*(3–4), 175–184. https://doi.org/10.1080/22797254.2021.1872929

Kreiterling, C. (2023). Digital innovation and entrepreneurship: A review of challenges in competitive markets. *Journal of Innovation and Entrepreneurship, 12*, 49. https://doi.org/10.1186/s13731-023-00320-0

Mazla, M. I. S. B., Jabor, M. K. B., Tufail, K., Yakim, A. F. N., & Zainal, H. (2020). The roles of creativity and innovation in entrepreneurship. In: *Proceedings of the International Conference on Student Disable Student Development 2019 (ICoSD 2019)*, pp 213–217. Atlantis Press. https://doi.org/10.2991/assehr.k.200921.035

Ruggieri, R., Savastano, M., Scalingi, A., Bala, D., & D'Ascenzo, F. (2018). The impact of digital platforms on business models: An empirical investigation on innovative start-ups. *Manag Mark, 13*(4), 1210–1225. https://doi.org/10.2478/mmcks-2018-0032

Szirmai, A., Naudé, W., & Goedhuys, M. (2011). Entrepreneurship, innovation, and economic development: An overview. In: Szirmai, A., Naudé, W., Goedhuys, M. (Eds.) *Entrepreneurship, innovation, and economic development*, pp. 1–20. Oxford: Oxford University Press. https://doi.org/10.1093/acprof:oso/9780199596515.003.0001

Veeraraghavan, V. (2009). Entrepreneurship and innovation. *Asia Pacific Business Review, 5*(1), 14–20. https://doi.org/10.1177/097324700900500102

Navigating Social Sustainability in India's Automobile Industry: A Data-Driven Approach to Achieving SDGs

Priyanka Singh, Pallavi Rajain, Chanchal Phore, and Chetna Grewal

Abstract Nowadays, sustainability has emerged as an essential component when assessing organizational practices, and our sector, the automobile industry, is under increased scrutiny due to its tremendous environmental and social impacts. The main objective of this research is to systematically identify and evaluate the social sustainability indicators that are most pertinent to the Indian automobile industry and emphasize how social sustainability contributes to the well-being of society throughout decades. The researcher used principal component analysis (PCA) to find critical social sustainability indicators in the Indian automobile industry. The outcomes indicated that the most significant determinants were gender diversity and workplace safety (LTIFR), followed by retention rate and board diversity. A dataset of 17 automobile companies listed on BSE and NSE in 2024 is used to provide insights for enhancing social sustainability practices. The study offers the Indian automobile industry useful advice on how to strengthen its social sustainability efforts by emphasizing on the most significant indicators. By giving these high-impact metrics a greater significance, these companies may enhance their brand values and bolster their contribution to the UN Sustainable Development Goals.

Keywords Sustainable development · Environment · Social · Automobile · Principal component analysis · Indicators

---

P. Singh (✉) · C. Grewal
Department of Commerce, Maharaja Surajmal Institute, Janakpuri, New Delhi, India
e-mail: priyanka@msijanakpuri.com

C. Grewal
e-mail: chetnagrewal@msijanakpuri.com

P. Rajain · C. Phore
Department of Management, Maharaja Surajmal Institute, Janakpuri, New Delhi, India
e-mail: pallavirajain@msijanakpuri.com

C. Phore
e-mail: chanchalphore@msijanakpuri.com

R. S. Mor et al. (eds.), *Sustainable Business Management, Innovation and Technology*, Sustainable Economy and Ecotechnology, https://doi.org/10.1007/978-3-031-98141-8_5

1 Introduction

The notion of a sustainable business model, or SBM for short, outlines the rationale for how an organization creates, offers, and acquires value in social, cultural, economic, and other contexts in an eco-friendly way. The process of developing an SBM is one innovative aspect of a business plan. Numerous companies and sectors have embraced the notion of SBM to simultaneously accomplish their social, environmental, and economic goals (Nosratabadi et al., 2019). As a component of the 2030 Agenda for Sustainable Development, which lays out a 15-year plan to achieve the Goals and their related targets, the Goals were accepted by all United Nations Member States in September 2015. World leaders had never before promised coordinated action on such a wide-ranging and global policy agenda. (United Nations, n.d.) Governments are expected to take charge and provide national frameworks for the accomplishment of the 17 Goals, even though the SDGs are not legally mandatory. All relevant parties, including governments, businesses, civil society, the UN system, and others, must work together to achieve the 17 Goals, which are interrelated and applicable to all nations. (United Nations, n.d.) The increase in global population hastened economic progress by increasing the demand for resources necessary for commercial and industrial operations, as well as having an impact on the environment (Bocken et al., 2014). Corporate pressure to be sustainable has also intensified as a result of this need (Kolk, 2016). Naturally, as concern about sustainability issues grows, businesses must show that they are rationally positioned on sustainability (Tseng et al., 2019), which incorporates social, environmental, and economic goals in addition to traditional commercial ones (Gimenez et al., 2012). By sharing information, technology, expertise, and costs, collaboration makes it possible to achieve sustainable, economic, environmental, and social goals while also boosting economic performance and competitiveness (Ray & Mondal, 2017). When developing strategies to advance in sustainability, organizations can better position their enterprises for survival by adopting a sustainable approach. To modernize its SBM (sustainable business model), it has to incorporate the three elements of the sustainable perspective. Economic techniques that incorporate environmental control strategies also take an environmental viewpoint. On the contrary, social strategy alludes to the investments and assistance provided in matters affecting customers, staff, and the community; these efforts might not be necessarily directly linked to business, but they nonetheless enhance the company's reputation (Wichaisri & Sopadang, 2017).

This study was carried out in the automobile industry. The automobile sector was chosen because the environmental impacts of the automobile sector are more easily isolated than the other industries like finance, mining, healthcare etc. The Indian automobile market is third in the world in term of sales. The contribution of automobile industry in the GDP is around 7.1%. India's auto industry is worth US$100 billion providing jobs to over 19 million people as of April 2022 (Banerjee, 2023). For present analysis, the following car manufacturers were chosen: Ashok

Leyland, Apollo Tyres, Bajaj, Balakrishna tyres, Bosch, Eicher motors, Escorts-Kubota, Force motors, Hero motors, Mahindra and Mahindra, Maruti, Motherson Samvardhana, MRF, SML Isuzu, Tata motors, Tube investments and TVS.

The structure of this paper is as follows: Sect. 2 presents a comprehensive review of the literature on sustainability, with a particular emphasis on the automobile sector; Sect. 3 details the methodology employed in this study; Sect. 4 provides a discussion of the results; and Sect. 5 concludes with a summary of the findings and their implications.

2 Review of Literature

2.1 Sustainability

The ability of an organization, procedure, or activity to be sustained over time without deteriorating or depleting resources or having a detrimental influence on the environment or society is known as sustainability (Johnston et al., 2007; Moore et al., 2017). This idea is frequently used in relation to the usage of natural resources, including minerals, water & land, as well as the effects of human activity on the environment (Peñarroya-Farell et al., 2023). The economic, environmental & social dimensions—also referred to as the three pillars—make up the societal goal of sustainability. Global, national, and individual consumer decisions can all be made with this idea in mind. The idea of sustainable development is connected to this one. Both words are frequently used interchangeably. UNESCO developed the following distinction: "Sustainability is often thought of as a long-term goal (i.e. a more sustainable world), while sustainable development refers to the many processes and pathways to achieve it." Sustainability at the corporate level in the business world generally involves striking a balance among a company's monetary growth, social welfare for all of its stakeholders, and preservation of the environment to meet the requirements of the present "without compromising the ability of future generations to meet their own needs." (Tolkamp et al., 2018).

Within the field of business innovation, sustainability is becoming more and more popular. Sustainable business strategies and efforts are becoming more prevalent. Many companies are putting in place environmentally conscious practices, like minimizing the emissions of carbon, reducing carbon emissions and contributing back to their communities (Chen et al., 2011; Menozzi et al., 2015). Significant advancements have also been made in sustainable industries: Green building and renewable energy are two sectors of the economy that are expanding significantly (Bryant et al., 2018; Chan & Lau, 2002). Additionally, customers want businesses to run more sustainably, which is driving up demand for socially and environmentally conscious goods as well as the popularity of consumer brands with a sustainability focus (Liu et al., 2018; Wu et al., 2021). Lastly, an analysis of the functions of governmental and regulatory organizations reveals that sustainability is a top policy objective and is probably going to stay popular in the business sector (Bryant et al., 2018).

2.2 Automobile Industry

The sustainable technology that we emphasize in this study, electric vehicles (EVs), additionally confronts the task of how to provide added benefits to customers, particularly to make up for their more expensive initial cost when compared with traditional cars (Kley et al., 2011). These benefits may result from EVs in the following ways: they may assist in creating greater comprehensiveness in mobility solutions (Kley et al., 2011); they may hold energy in "smart energy" systems (Ceschin & Vezzoli, 2010); they can create novel revenue streams by leasing batteries (Christensen et al., 2012); and they can recycle batteries for applications that require another chance (Neubauer & Pesaran, 2011). It is understandable, though, that the burgeoning EV "industry" is still in need of a workable business plan (Christensen et al., 2012; Kley et al., 2011). As Teece (2010) notes, "the right business model is rarely apparent early on in emerging industries." Through a process of learning, experimenting & adaptation, several business models are being explored simultaneously both across and within organizations (Demil & Lecocq, 2010; Sosna et al., 2010; Teece, 2010).

Tesla is an automobile firm that aims to revolutionize the market by providing electric vehicles (EVs) at a competitive price. At the strategic level, the firm hopes to support global efforts to transition to sustainable energy through its goal statement, which is, "to accelerate the advent of sustainable transport by bringing compelling mass-market electric cars to market as soon as possible" (Rowland, 2016). Tesla intends to accomplish this goal by employing its electric motor technology (Martins et al., 2015). The company began by providing the market with a sustainable value proposition that included lithium-ion battery packs and battery electric vehicles (BEVs) that were efficient, emission-free, and environmentally beneficial. Later on, the business added to this value proposition by providing fast-charging infrastructure powered by solar energy in order to provide clean energy solutions. By utilizing innovative technology and compassionate leadership, the organization was able to take advantage of several business growth prospects. Originally servicing the premium sports vehicle market primarily EVs, the company has broadened its value proposition by offering a range of BEV (battery electric vehicle) ranges for additional customer groups (Hernández-Chea et al., 2021). The business was able to steadily increase revenue (Dyer et al., 2020) and improve its standing in the international marketplace.

The business now sells cars directly to consumers rather than through a franchised dealership, forging a direct contact with B2C clients in the process. In order to deliver customers with a distinctive shopping experience, Tesla built a direct relationship with them. The company also continuously engaged with them to get input about the performance of the vehicles. Through such partnerships, the company was able to affect its partners' and collaborators' sustainable business practices while also contributing to the buying patterns and perspectives of its clientele about the personal use of electric vehicles (Hernández-Chea et al., 2021).

3 Methodology

3.1 Objective

To systematically identify and evaluate the social sustainability indicators that are most pertinent to the Indian automobile industry.

To analyze the sustainability performance of automobile companies concerning the selected Sustainable Development Goals (SDGs).

3.2 Methods

To analyse the objectives of this study the current global UN SDG indicators were used. For selection of relevant environmental SDG indicators, firstly, criteria need to be defined for characterizing the indicators as "relevant". Next, for these criteria it is specified which indicators are selected or not selected.

Data Collection Procedure for Social Sustainability Indicators from India's Automobile Sector

The dataset utilized in this study contains comprehensive information on the social sustainability indicators reported by automobile companies. Seventeen automobile companies listed in the Bombay Stock Exchange (BSE) and National Stock Exchange (NSE) in the year 2024 were included in our dataset Although we aimed to consider all the listed automobile companies, some such as Sundaram Fasteners and Cummins India did not provide any information about their sustainability practices and outcomes in their annual reports. Therefore, our final dataset consisted of 17 companies: Ashok Leyland, Apollo Tyres, Bajaj Auto, Balakrishna Industries, Bosch, Eicher Motors, Escorts Kubota, Hero Motors, Mahindra & Mahindra, Maruti Suzuki, Motherson Samvardhana, MRF Tyres, SML Isuzu, Tata Motors, Tube Investments, TVS Motors and Force Motors. We identified over 15 social indicators which are relevant to the automobile and automotive sector; however, a careful examination of the above companies' BSBR (business responsibility and sustainability report) and annual reports showed that many indicators' data was not reported. For instance, none of the companies reported their R&D investments in sustainable technology development, work culture, or investment in education. We therefore selected only those social sustainability indicators which were available in the BSBR and annual reports of most companies, represented in Table 1.

Table 1 Data variables classification

Social sustainability indicators
1. Hired
2. Gender diversity
3. LTIFR
4. Retention rate
5. Board diversity
6. Total trained employees/workers
7. Independent directors in board
8. Total wages (in million)

[a]*Author's compilation*

3.3 Data Analysis

3.4 Principal Component Analysis for Social Sustainability in India's Automobile Sector

Leading automobile companies are adopting strategies that meet UN sustainable development goals and contribute not only towards global sustainability but also to strengthening the brand value of the companies (Lukin et al., 2022). Previous literature has shown that sustainability performance of a company can broadly be judged by two major categories of indicators- Environmental Indicators: These measure the environmental impact of a company's operations, including emissions, energy consumption, water usage, CO_2 abatement, and waste management. Social Indicators: These reflect aspects of employee well-being, including workplace safety, diversity, community engagement, and overall workforce satisfaction (Ly & Cope, 2023). The aim is to identify which of the many social sustainability indicators are relevant to the Indian automobile industry. The purpose is to help the automobile sector improve its social sustainability practices by focusing specifically on the relevant indicators. This will help them not only achieve their sustainability goals in a simpler and more efficient way, but also improve their social performance.

3.5 The Purpose of Dimension Reduction of Social Sustainability Indicators

While these indicators provide a comprehensive framework for sustainability reporting, it is unclear whether they are all equally effective in measuring their respective dimensions of sustainability. Some indicators may contribute significantly to defining environmental or social sustainability, while others might add little value or introduce redundancy. For instance, Lisowski et al. (2020) initially identified

247 indicators of environmental sustainability in the automobile sector, of which only 31 were found to be relevant. Similarly, Li et al. (2012) found that out of 32 sustainability indicators, only 13 were found to be relevant after empirical examination of sustainability reports of electronics companies. Dimensionality reduction is important because there is a "practical difficulty in working with a large number of metrics, and the goal is to reduce it to a more manageable set of core indicators addressing many different areas" (Munier, 2011). Moreover, sustainability is a complex phenomenon, and to capture its complexity, composite indices must be created by integrating multiple indicators (Gan et al., 2017). Therefore, we propose to apply the PCA test to reduce the number of social sustainability indicators from 8 to 3 or 4 which accurately represent the companies' social sustainability performance.

3.6 Principal Component Analysis (PCA) to Extract Relevant Variables

PCA is a technique which helps in dimension reduction and better interpretation of large datasets (Jolliffe & Cadima, 2016). Several past studies have advocated the use of PCA test for aggregating sustainability indicators. For instance, Zhao (2015) applied the PCA test to extract principal components to represent nine indices covering four aspects of a country's sustainability. Martins et al. (2021) performed a neighbourhood-level assessment of urban sustainability for different regions in Brazil and identified 14 social sustainability indicators. Finally, Cao et al. (2023) used PCA to reduce 13 indicators of higher education quality and sustainability of educational institutions to four principal components. Therefore, this study too aims to enhance the precision and practicality of sustainability evaluations by using PCA to reduce these 9 sustainability indicators to fewer principal components. This will support automobile companies in achieving more transparent and impactful sustainability practices.

3.7 Data Preparation

Raw data collected from secondary reports are not always appropriate for making deductions about the data (Amiram et al., 2015; Crucean, 2024; Sabauri & Kvatashidze, 2024; Turkmen, 2016). Therefore, there is a need to process it to have a better understanding of the data. Herein, the process adopted for making data ready for PCA analysis is as follows.

Step 1. Data Loading: Our initial dataset was a cohesive one comprising environmental and social sustainability indicators. This dataset was imported from an Excel file. The data was read into a pandas Dataframe by connecting the Google Colab Jupyter notebook with Google Drive.

Step 2. Data Modification: Before proceeding with PCA analysis, we want to ensure that all the data is numerical. So, this can be done by identifying the datatypes. The. D types attribute will return a Series with the data types of each column. Herein, scope 3 variable is object type so we need to convert it into float64 only. For this, firstly we need to convert the information to numeric and handle all the error values. pd.to_numeric(df['Scope 3'], errors = 'coerce') function attempts to convert values in column 'Scope 3' to numeric and df['Scope 3'].astype('float64') the variable value is converted to floadt64 type (as shown in Fig. 1).

Step 3. Splitting the Data into 'Social' and 'Environmental' DataFrames.

Given that the focus is on social indicators, the data was split into two DataFrames-Social DataFrame: Contains columns related to social indicators as specified in Fig. 2. Environmental DataFrame: Contains columns related to environmental indicators as mentioned in Fig. 2.

Scope 1	float64
Scope 2	float64
Scope 3	object
Energy Consumed (GJ)	float64
% share of renewable electricity	float64
Total Fuel Consumption (GJ)	float64
Total Electricity Consumed (GJ)	float64
Total Renewable Electricity (GJ)	float64
CO2 sequestered through afforestation	float64
through renewable energy	float64
through eneregy saving projects	float64
trees planted	float64
Total water consumed (KL)	float64
Water recycled	float64
Hired	float64
Gender Diversity	float64
LTIFR	float64
Retention rate	float64
Board diversity	float64
Total trained employees/ workers	int64
Independent Directors in Board	float64
Total wages (in million)	float64
Waste Generated	float64
Waste Disposed	float64
Waste Recycled/Reused	float64

Fig. 1 Variables

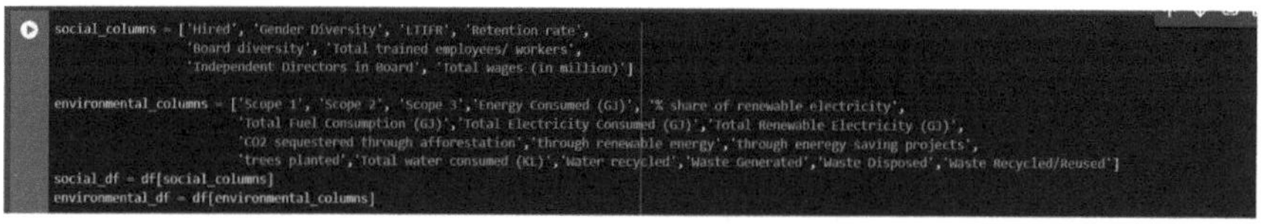

Fig. 2 Data splitting

Step 4. Data Cleaning: PCA requires that missing values be handled, as it cannot be performed on datasets with NaN values. So, missing values need to be eliminated using dropna() method. However, when we eliminated values we could see that only 5 observations are left because data for hired, independent directors in board and total wages was not available. This reduces the efficiency of the data. So, to handle this what we do is remove all these indicators from the model. Using the drop() Method we can specify the columns we want to remove by passing their names to the drop() method. Now, to account for any other missing values in the dataset, dropna() method will be used. Finally missing values are removed and made the dataframe relevant for PCA processing.

4 Discussion and Insights on Social Sustainability in the Automobile Sector

As we want to know whether the specified indicators are relevant to measuring the social sustainability aspect or not, we specify the number of components to 1. Table 2 highlights key social sustainability indicators in India's automobile sector and their alignment with relevant SDGs.

Table 2 Social sustainability indicators and their alignment with SDGs

Indicator	Description	SDG Alignment
Gender diversity	Percentage of female employees in the workforce	SDG 5 (gender equality)
LTIFR (workplace safety)	Lost time injury frequency rate (LTIFR)—measures workplace safety	SDG 8 (decent work & economic growth)
Retention rate	Employee retention percentage over time	SDG 8 (decent work & economic growth)
Board diversity	Percentage of independent and female directors on the board	SDG 10 (reduced inequalities)
Total trained employees	Number of employees receiving formal training programs	SDG 4 (quality education)

[a] *Author's compilation*

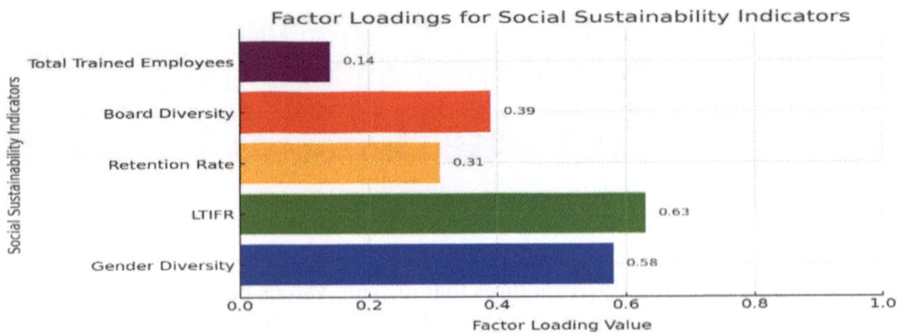

Fig. 3 Indicator wise factor loading

4.1 PCA Model Fitting

With this specification, the PCA model is fitted on normalized data and the results are derived. The PCA applied data frame shows that now the dimension of the data frame has reduced from 14 observations and 5 columns to 14 observations and 1 dimension i.e. social sustainability. The explained variance ratio of the first principal component (PC1) is **33.67%**. This means that the 1 component developed for representing the social sustainability explained a substantial portion (33.67%) of the total variance in indicators. By focusing on one component, we capture the most significant trends without adding complexity. A scree plot was generated to display the explained variance for each component. Finally, loadings are derived to reveal the contribution of each indicator to the social sustainability component.

Figure 3 shows that Gender diversity has a high factor loading of 0.58, indicating a strong contribution to the principal component. This suggests that gender diversity is a key determinant of social sustainability in the dataset.

LTIFR has the highest factor loading i.e. 0.63, meaning it has the strongest influence on the principal component. This implies that workplace safety, as measured by LTIFR, is critical in defining social sustainability. The retention rate has a moderate loading of 0.31, showing a noticeable but less significant contribution to the principal component. It plays a role in social sustainability but is not as impactful as gender diversity or LTIFR. Board diversity also has a moderate contribution of 0.39, slightly higher than the retention rate. This indicates that diversity in leadership impacts social sustainability but is secondary to workplace safety and gender diversity. Total Trained Employees/Workers has the lowest factor loading i.e. 0.14, suggesting a minimal contribution to the principal component. While training is relevant, it is not as impactful in defining social sustainability in this dataset. Based on the trend in factor loading, we have selected approx. 0.3 as the criteria for identifying the most relevant indicators for measuring the social sustainability indicator (Gorsuch, 2014). The indicators with the most substantial loadings, contributing significantly to the social sustainability score, are gender diversity, LTIFR, retention rate, and board diversity. These factors are crucial for achieving high social sustainability performance.

5 Conclusion and Implications

Sustainability has become a fundamental parameter for evaluating organizational practices, especially within industries like automobile, where environmental and social factors are gaining increased attention. This study focuses on social sustainability, a critical component for fostering societal well-being across generations. The analysis highlights that workplace safety (LTIFR) and gender diversity are the most influential indicators of social sustainability, followed by retention rate and board diversity, which also play important roles. In contrast, the contribution of employee training metrics is relatively minimal in this dataset, suggesting the need for further exploration to understand their impact. Past research too has shown that LTIFR is an important social sustainability indicator, as it represents occupational health and safety and hazard prevention techniques used at the workplace (Koskela, 2014), while (Arayakarnkul et al., 2022; Galdiero et al., 2024; Grosser, 2009; Lin & Efranto, 2023) find gender diversity is an important contributor to the broader sustainability goals of the United Nations.

The findings emphasize the importance of aligning organizational strategies with these critical indicators to improve social sustainability performance. By focusing on these high-impact metrics, automobile companies can enhance their contribution to the UN Sustainable Development Goals while simultaneously strengthening their brand value. The application of PCA not only simplifies the analysis of complex sustainability datasets but also ensures precision in identifying impactful variables. As the automobile sector continues to adopt sustainable practices, incorporating the identified indicators into strategy and reporting frameworks will be pivotal for fostering long-term social and environmental well-being.

References

Amiram, D., Bozanic, Z., & Rouen, E. (2015). Financial statement errors: Evidence from the distributional properties of financial statement numbers. *Review of Accounting Studies, 20*, 1540–1593. https://doi.org/10.1007/s11142-015-9333-z

Arayakarnkul, P., Chatjuthamard, P., & Treepongkaruna, S. (2022). Board gender diversity, corporate social commitment and sustainability. *Corporate Social Responsibility and Environmental Management, 29*, 1706–1721. https://doi.org/10.1002/csr.2320

Banerjee, A. (2023). *Economic survey 2023: Auto sector to play a critical role in India's shift to green energy*. https://www.moneycontrol.com/news/business/economic-survey-2023-auto-sector-to-play-a-critical-role-in-indias-shift-to-green-energy-9971981.html

Bocken, N. M., Short, S. W., Rana, P., & Evans, S. (2014). A literature and practice review to develop sustainable business model archetypes. *Journal of Cleaner Production, 65*, 42–56.

Bryant, G. (2018). Nature as accumulation strategy? Finance, nature, and value in carbon markets. *Annals of the American Association of Geographers, 108*(3), 605–619.

Cao, C., Wei, T., Xu, S., Su, F., & Fang, H. (2023). Comprehensive evaluation of higher education systems using indicators: PCA and EWM methods. *Humanities & Social Sciences Communications, 10*, 1–12. https://doi.org/10.1057/s41599-023-01938-x

Ceschin, F., Vezzoli, C., & Zhang, J. (2010). *Sustainability in design: Now! Challenges and opportunities for design research, education and practice in the XXI century.*

Chan, R. Y., & Lau, L. B. (2002). The effectiveness of environmental claims among Chinese consumers: Influences of claim type, country disposition and ecocentric orientation. *Journal of Marketing Management, 20*(3–4), 273–319.

Chen, I. C., Hill, J. K., Ohlemüller, R., Roy, D. B., & Thomas, C. D. (2011). Rapid range shifts of species associated with high levels of climate warming. *Science, 333*(6045), 1024–1026.

Christensen, T. B., Wells, P., & Cipcigan, L. (2012). Can innovative business models overcome resistance to electric vehicles? Better Place and battery electric cars in Denmark. *Energy Policy, 48*, 498–505.

Crucean, A. (2024). (PDF) The influence of the risk of fraud and accounting errors in the financial statements on the opinion of the financial auditor. *The European Journal of Finance Business, 7*, 2019. https://doi.org/10.4316/EJAFB.2019.711

Demil, B., & Lecocq, X. (2010). Business model evolution: In search of dynamic consistency. *Long Range Planning, 43*(2–3), 227–246.

Dyer, J., Furr, N., & Hendron, M. (2020). Overcoming the innovator's paradox. *MIT Sloan Management Review, 61*(4), 1–9.

Galdiero, C., Maltempo, C., Marrapodi, R., & Martinez, M. (2024). Gender diversity: An opportunity for socially inclusive human resource management policies for organizational sustainability. *Social Sciences, 13*, 173. https://doi.org/10.3390/socsci13030173

Gan, X., Fernandez, I. C., Guo, J., Wilson, M., Zhao, Y., Zhou, B., & Wu, J. (2017). When to use what: Methods for weighting and aggregating sustainability indicators. *Ecological Indicators, 81*, 491–502. https://doi.org/10.1016/j.ecolind.2017.05.068

Gimenez, C., Sierra, V., & Rodon, J. (2012). Sustainable operations: Their impact on the triple bottom line. *International Journal of Production Economics, 140*(1), 149–159.

Gorsuch, R. L. (2014). *Factor analysis* (Classic). Taylor & Francis.

Grosser, K. (2009). Corporate social responsibility and gender equality: Women as stakeholders and the European Union sustainability strategy. *Business Ethics: A European Review, 18*, 290–307. https://doi.org/10.1111/j.1467-8608.2009.01564.x

Hernández-Chea, R., Jain, A., Bocken, N. M., & Gurtoo, A. (2021). The business model in sustainability transitions: A conceptualization. *Sustainability, 13*(11), 5763.

Johnston, P., Everard, M., Santillo, D., & Robèrt, K. H. (2007). Reclaiming the definition of sustainability. *Environmental Science and Pollution Research International, 14*(1), 60–66.

Jolliffe, I. T., & Cadima, J. (2016). Principal component analysis: A review and recent developments. *Philosophical Transactions of the Royal Society a Mathematical Physical and Engineering Sciences, 374*(2065), 20150202. https://doi.org/10.1098/rsta.2015.0202

Kley, F., Lerch, C., & Dallinger, D. (2011). New business models for electric cars—A holistic approach. *Energy Policy, 39*(6), 3392–3403.

Kolk, A. (2016). The social responsibility of international business: From ethics and the environment to CSR and sustainable development. *Journal of World Business, 51*(1), 23–34.

Koskela, M. (2014). Occupational health and safety in corporate social responsibility reports. *Safety Science, 68*, 294–308. https://doi.org/10.1016/j.ssci.2014.04.011

Li, T., Zhang, H., Yuan, C., Liu, Z., & Fan, C. (2012). A PCA-based method for construction of composite sustainability indicators. *The International Journal of Life Cycle Assessment, 17*(5), 593–603.

Lin, C. J., & Efranto, R. Y. (2023). Do age and gender change the perception of workplace social sustainability? *Sustainability, 15*, 5013. https://doi.org/10.3390/su15065013

Liu, J., Hull, V., Godfray, H. C. J., Tilman, D., Gleick, P., Hoff, H., Pahl-Wostl, C., Xu, Z., Chung, M.G., Sun, J. & Li, S. (2018). Nexus approaches to global sustainable development. *Nature sustainability, 1*(9), 466–476

Lisowski, S., Berger, M., Caspers, J., Mayr-Rauch, K., Bäuml, G., & Finkbeiner, M. (2020). Criteria-based approach to select relevant environmental SDG indicators for the automobile industry. *Sustainability, 12*, 8811. https://doi.org/10.3390/su12218811

Lukin, E., Krajnović, A., & Bosna, J. (2022). Sustainability strategies and achieving SDGs: A comparative analysis of leading companies in the automotive industry. *Sustainability, 14*(7), 4000. https://doi.org/10.3390/su14074000

Ly, A. M., & Cope, M. R. (2023). New conceptual model of social sustainability: Review from past concepts and ideas. *International Journal of Environmental Research and Public Health, 20*(7), 5350. https://doi.org/10.3390/ijerph20075350

Martins, L. L., Rindova, V. P., & Greenbaum, B. E. (2015). Unlocking the hidden value of concepts: A cognitive approach to business model innovation. *Strategic Entrepreneurship Journal, 9*(1), 99–117.

Martins, M. S., Kalil, R. M. L., & Rosa, F. D. (2021). Sustainable neighbourhoods: Applicable indicators through principal component analysis. *Proceedings of the Institution of Civil Engineers-Urban Design and Planning, 174*, 25–36. https://doi.org/10.1680/jurdp.20.00058

Menozzi, D., Fioravanzi, M., & Donati, M. (2015). Farmer's motivation to adopt sustainable agricultural practices. *Bio-Based and Applied Economics, 4*(2), 125–147.

Moore, F. C., Baldos, U., Hertel, T., & Diaz, D. (2017). New science of climate change impacts on agriculture implies higher social cost of carbon. *Nature Communications, 8*(1), 1607.

Munier, N. (2011). Methodology to select a set of urban sustainability indicators to measure the state of the city, and performance assessment. *Ecological Indicators, 11*, 1020–1026. https://doi.org/10.1016/j.ecolind.2011.01.006

Neubauer, J., & Pesaran, A. (2011). The ability of battery second use strategies to impact plug-in electric vehicle prices and serve utility energy storage applications. *Journal of Power Sources, 196*(23), 10351–10358.

Nosratabadi, S., Mosavi, A., Shamshirband, S., Zavadskas, E. K., Rakotonirainy, A., & Chau, K. W. (2019). Sustainable business models: A review. *Sustainability, 11*(6), 1663.

Peñarroya-Farell, M., Miralles, F., & Vaziri, M. (2023). Open and sustainable business model innovation: An intention-based perspective from the Spanish cultural firms. *Journal of Open Innovation: Technology, Market, and Complexity, 9*(2), Article 100036.

Ray, A., & Mondal, S. (2017). Study of collaborative PRM business model for sustainability. *Benchmarking: an International Journal, 24*(7), 1891–1911.

Rowland, C. Tesla, Inc.'s Mission Statement & Vision Statement (An Analysis). 2016. Available online: http://panmore.com/tesla-motors-inc-vision-statement-mission-statement-analysis

Sabauri, L., & Kvatashidze, N. (2024). Sustainability reporting issues. *Journal of Entrepreneurship Sustainable*. https://doi.org/10.9770/jesi.2023.11.2(19)

Sosna, M., Trevinyo-Rodríguez, R. N., & Velamuri, S. R. (2010). Business model innovation through trial-and-error learning: The Naturhouse case. *Long Range Planning, 43*(2–3), 383–407.

Teece, D. J. (2010). Business models, business strategy and innovation. *Long Range Planning, 43*(2–3), 172–194.

Tolkamp, J. C. C. M., Huijben, J. C. C. M., Mourik, R. M., Verbong, G. P. J., & Bouwknegt, R. (2018). User-centred sustainable business model design: The case of energy efficiency services in the Netherlands. *Journal of Cleaner Production, 182*, 755–764.

Tseng, C. H., Chang, K. H., & Chen, H. W. (2019). Strategic orientation, environmental innovation capability, and environmental sustainability performance: The case of Taiwanese suppliers. *Sustainability, 11*(4), 1127.

Turkmen, B. (2016). Errors and Abuses in Financial Accounting and Results. *Procedia Econ. Finance*. https://doi.org/10.1016/S2212-5671(16)30179-4

United Nations. (n.d.-b). *Sustainable Development Goals | United Nations*. https://www.un.org/en/academic-impact/page/sustainable-development-goals

Wichaisri, S., & Sopadang, A. (2017). Integrating Liu, J., Hull, V., Godfray, H. C. J., Tilman, D., Gleick, P., Hoff, H., Pahl-Wostl, C., Xu, Z., Chung, M.G., Sun, J. & Li, S. (2018). Nexus approaches to global sustainable development. Nature sustainability, 1(9), 466-476. development, lean, and logistics concepts into a lean sustainable logistics model. *International Journal of Logistics Systems and Management, 26*(1), 85–104.

Wu, X., Fu, B., Wang, S., Song, S., Li, Y., Xu, Z., Wei, Y. & Liu, J. (2022). Decoupling of SDGs followed by re-coupling as sustainable development progresses. *Nature Sustainability, 5*(5), 452–459

Zhao, G. (2015). *A sustainability classification for a country based on PCA*. Atlantis Press, Baoding.

Understanding (KM) Knowledge Management with Philosophical Frameworks: A Study of Sustainability

Arunima, Sonal Jauhari, and Pratham Pratap Singh

Abstract Knowledge management (KM) is based on both an understanding of knowledge creation and knowledge implementation. Through philosophical frameworks humans are able to examine the underlying principles that guide how knowledge is created, shared, and utilized to promote sustainable practices. Utilitarianism focuses on actions that produce the greatest good for the greatest number. Deontology emphasizes on ensuring that (KM) knowledge management is aligned with moral obligations towards future generation and the environment. The following paper will highlight Constructivism as how (KM) Knowledge management works as co-creation; knowledge is based on social interactions and individual experience. Though this descriptive paper we wish to prove that sustainable development is possible when (KM) knowledge management is driven by philosophical frameworks and pragmatism. Emphasizing the practical consequences of ideas. This research paper explores the interplay between knowledge management and sustainability, viewed through the philosophical lens of utilitarianism, deontology, virtue ethics, constructivism, and epistemic justice. Integrating philosophical frameworks in (KM) knowledge management can create strategies that are ethically sound and can promote sustainable development. This paper will not only enhance organizational performance in particular but also contribute to broader societal and environmental goals as a whole.

Keywords Epistemic justice · Pragmatism · Constructivism · Philosophy · (KM) Knowledge management · Utilitarianism

Arunima (✉) · S. Jauhari · P. P. Singh
(DIHE) Delhi Institute of Higher Education, Jims Noida Extension, Greater Noida West, India
e-mail: arunima.dihe@dihe.co.in

S. Jauhari
e-mail: sonal.jauhari@dihe.co.in

P. P. Singh
e-mail: Prathamsingh.st@dihe.co.in

R. S. Mor et al. (eds.), *Sustainable Business Management, Innovation and Technology*, Sustainable Economy and Ecotechnology, https://doi.org/10.1007/978-3-031-98141-8_6

1 Introduction

In today's dynamic and knowledge-driven environment, organizations rely on more than just information and resources to achieve sustainable success; they must also foster an innovative culture that embraces adaptability, ethical considerations, and effective leadership. Knowledge Management (KM) has emerged as a critical discipline aimed at enhancing organizational capability by creating, sharing, and utilizing knowledge effectively. For KM practices to be truly effective and sustainable, they must be viewed not only as operational strategies but also through a philosophical lens. This approach integrates ethical principles, deepens understanding of knowledge itself, and links leadership values with organizational sustainability, fostering a more comprehensive view of KM.

Philosophy provides a foundational framework for examining the nature of knowledge, the ethics that govern its use, and the influence of human values and leadership on KM practices. By adopting a philosophical perspective, organizations can explore KM beyond functionality, aligning it with ethical considerations and sustainability goals. Such a framework is especially pertinent in an age where stakeholders—employees, customers, and the public—expect organizations to uphold ethical standards, social responsibility, and environmental consciousness.

This research delves into the synergies between philosophy and KM to explore three core objectives. First, Assesses the Role of Philosophical Knowledge Management (KM) Approaches in Promoting Positive Thinking and Adaptability in Organizations considering how a philosophically grounded approach can empower organizations to navigate complex and changing environments. Second, it investigates how ethical considerations in KM practices affect employee engagement and retention, recognizing that when knowledge practices are ethical, they resonate more deeply with employees, cultivating trust and loyalty. Lastly, the study examines the influence of leadership on the sustainability of KM practices, recognizing that visionary leadership rooted in ethical and philosophical principles is instrumental in embedding KM as a lasting component of organizational culture. By addressing these objectives, this paper aims to elucidate the critical intersections between KM, philosophy, and sustainability. It demonstrates how philosophical insights into knowledge and ethics can shape KM practices that are not only innovative but also ethically sound and aligned with long-term organizational sustainability. Through this approach, the study aims to delve into the significant role that philosophical principles play in shaping knowledge management (KM) practices within organizations keeping in mind the ethical considerations and practices. It focuses on how these principles foster innovation, promote ethical behaviour, and contribute to long-term sustainability in that particular organization.

1.1 Understanding Philosophy

Philosophy the word comes from the Greek word "Philo" meaning Love and "Sophia" meaning Wisdom. Wisdom is the highest form of thinking and cognitive ability which facilitates human being to discriminate between right and wrong. It also helps in taking right and sound decisions for self and others as a whole. Philosophy can help the five highly interrelated functions of knowledge that could aid in fulfilling the tasks of any individual: (i) enabling the individual to resolve dilemmas and make decisions in her/his life, (ii) advising others, (iii) Managing and guiding society, (iv) Reviving life and (v) questioning of life's meaning. **Knowledge** is a fundamental driver of success, growth, and innovation, and **Knowledge Management** provides the processes and structures to capture, share, and apply this knowledge effectively. Together, they form the backbone of intellectual capital and intelligent decision-making. It facilitates organizational development, and competitive advantage in today's world. It is rightly said:

> Knowledge is a fluid mu of framed experience, values, contextual information, and expert insight that provides a framework for evaluating and incorporating the new experiences and information It originates and is applied in the minds of knowers. In organizations, it often becomes embedded not only in documents and repositories but also in organizational routines, processes, practices and norms. IS literature. First, [8, p.31]

1.2 Knowledge Management

Knowledge Management (KM) refers to the process of identifying, capturing, organizing, sharing, and utilizing knowledge to enhance the effectiveness and performance of an organization. It encompasses strategies and practices that help organizations create, store, and apply knowledge in ways that foster innovation, improve decision-making, and achieve strategic objectives. KM integrates both tacit knowledge (personal, experience-based knowledge) and explicit knowledge (documented, codified information) to optimize organizational learning and improve business outcomes. Alavi and Leidner (2001) in their paper *Knowledge Management and Knowledge Management Systems: Conceptual Foundations and Research Issues* provide a more technical view of KM-"Knowledge management refers to a set of practices aimed at improving the creation, sharing, and utilization of knowledge in organizations."-Alavi and Leidner (2001).

1.3 Philosophy and Knowledge Management

Philosophy and **knowledge management (KM)** is both deep and insightful, as philosophical concepts can help shape, define, and guide the practices and principles of knowledge management. Philosophy, especially in areas like epistemology (the theory of knowledge), ethics, and ontology (the study of being), provides foundational insights that help in understanding how knowledge is created, shared, and utilized within organizations. Epistemology, the branch of philosophy that studies the nature, sources, and limits of knowledge, has a direct impact on how **knowledge management** is understood and practiced.

The fact that we need to get in-depth understanding of knowledge being shared to us. The true comprehension of knowledge will be possible only when we are able to understand the reason for its existence. **Ontology**, as a branch of philosophy, is concerned with the study of existence, being, and reality. Ontology explores what it means for something to "exist." This involves discussions about the nature of being itself and what kinds of things are said to exist. Do abstract objects like numbers or propositions exist in the same way as concrete physical objects like trees or mountains? The study of ontology began with Aristotle in the fourth century BCE. In his work Metaphysics, Aristotle investigates the nature of being, defining it as "that which is." He introduces key concepts such as substance, accidents, and categories—different ways in which beings can be categorized. For Aristotle, the fundamental "substance" underlies all properties and changes in the world. Martin Heidegger, a key figure in existential and phenomenological philosophy, redefined ontology in terms of human existence (Dasein). Heidegger argued that the way humans engage with the world shapes our understanding of being itself. He contended that traditional ontology had neglected the human experience of being, and instead, he focused on themes like time, death, and authenticity.

The link between **philosophy** and **ontology in Knowledge Management (KM)** can be seen in how ontological principles—rooted in philosophical traditions—inform the way knowledge is conceptualized, structured, and managed within organizations. Understanding the philosophical foundations of ontology can offer deep insights into how knowledge is classified, interpreted, and shared in a KM system. Ontology, in its philosophical sense, concerns the study of being and existence. Key philosophical traditions that influence the concept of ontology in knowledge management include: **Aristotelian Ontology**: Aristotle's work on categories and substance theory has directly impacted how knowledge is classified and structured. For example, Aristotle's categories (substance, quantity, quality, relation, etc.) can be applied to categorize knowledge in KM systems. In this sense, an organization's knowledge might be classified into categories like "procedural knowledge," "tacit knowledge," and "explicit knowledge." **Kantian Ontology**: Kant's philosophy of knowledge, where he discusses how knowledge is shaped by the interaction between sensory experience and mental categories, aligns with KM in terms of how organizational knowledge is shaped by the frameworks and categories individuals use to make sense of information. **Heideggerian Ontology**: Martin Heidegger's existential ontology, focusing on human experience and the "being-in-the-world" (Dasein), offers an understanding of how knowledge is not just an abstract collection of facts but is deeply connected to human experience, action, and interpretation. This idea can influence how knowledge management systems are designed to account for the lived experience of knowledge workers, especially in terms of tacit knowledge.

1.4 Pragmatism: Knowledge as a Tool for Action

Pragmatism is a philosophical tradition that views knowledge not as an abstract, detached entity but as a **tool for practical action.** According to pragmatism, knowledge's value is determined by its ability to solve problems and guide action effectively in the real world. Key thinkers such as **Charles Sanders Peirce, William James,** and **John Dewey** emphasized that knowledge is only valuable if it helps people address their everyday problems and adapt to changing circumstances. **Practical Consequences**: Knowledge should be evaluated based on its **practical utility.** Truth, in the pragmatist view, is what works in practice and produces beneficial outcomes. In the pragmatist tradition, knowledge serves as a tool for guiding action. It's evaluated by how well it helps individuals or groups **act effectively** in different contexts.

1.5 Constructivism

Constructivism posits that knowledge is actively **constructed** by individuals or groups through their experiences, social interactions, and engagement with the world. This philosophical view challenges the idea that knowledge is merely "discovered" or passed from one person to another; instead, it is seen as something created through human activity, particularly in social contexts. **Jean Piaget** and **Lev Vygotsky** are central figures in the development of constructivist theory. **Active Knowledge Construction**: Learners are not passive recipients of knowledge; they actively build knowledge through their experiences and interactions. **Contextual and Situated Knowledge**: Knowledge is always **context-dependent,** shaped by the environment, culture, and experiences of individuals. It evolves in response to ongoing interaction with the world. **Establishing Moral Principles** Philosophy helps ethical leaders identify and apply core moral principles that guide their behavior and decision-making. Some of the key philosophical ethical theories that influence leadership are: **Deontological Ethics** (e.g., Immanuel Kant): Focuses on duty and principles. Leaders guided by deontological ethics prioritize moral rules and duties, regardless of the outcomes. They understand that certain actions are inherently right or wrong, and leadership decisions should respect these principles (e.g., honesty, integrity, respect for human rights). **Utilitarianism** (e.g., John Stuart Mill, Jeremy Bentham):

Emphasizes the greatest good for the greatest number. Ethical leaders applying utilitarian principles make decisions that aim to benefit the most people, weighing the consequences of their actions to maximize overall well-being. **Virtue Ethics** (e.g., Aristotle): Focuses on the character of the leader. Virtue ethics suggests that ethical leadership is about cultivating good character traits (e.g., courage, honesty, fairness) and leading by example. A virtuous leader is one who acts with wisdom, temperance, and kindness. **Care Ethics** (e.g., Carol Gilligan): Stresses the importance of relationships and care for others. Leaders guided by care ethics prioritize empathy, compassion, and the well-being of individuals and communities over abstract principles. Leaders often face situations where they must make tough decisions that could conflict with personal or organizational interests. Philosophical thinking reinforces the importance of: **Integrity**: Leaders grounded in philosophy understand that being ethically consistent is crucial to maintaining trust. Integrity involves sticking to one's moral principles despite external pressure, whether from the organization, clients, or other stakeholders. **Moral Courage**: A philosophical understanding of ethics helps leaders to act on their beliefs, even when it involves personal sacrifice. Whether standing up for a marginalized group or resisting unethical practices within an organization, moral courage allows ethical leaders to take bold actions that uphold moral values. Philosophy plays a role in shaping the ethical culture within an organization.

1.6 Sustainability in Knowledge Management

Knowledge management is considered as the core competency, a primary and vernacular source of sustainability in management discipline for ethics and value creation for organizations all around the world. (Liu et al., 2018). Sustainability concern in combination with knowledge management has recently captured exceptional attention. From various researchers and practitioners. Focusing on the current market scenarios where the demand curve for knowledge is being considered and analyzed. Frequently leaning and indicating the significant demand at a certain price point for the amount of information, which is less readily accessible than what is socially preferred amount. Past studies explored that sustainability concept under various context of alignment in knowledge management in terms of longevity of knowledge practices and systems, resilience to withstand the challenges and adapt it overtime, continuity, endurance, perpetuity towards goals of maintaining knowledge resources and practices indefinitely. The value of knowledge is established by the extent of knowledge sharing within the organization. When knowledge sharing occurs with external partners it is subject to certain transfer restrictions due to the risk of industrial espionage (Brinkley, 2006). Sustainability in knowledge management (KM) is the practice of ensuring that knowledge process systems, and assets **remain** effective, relevant, and beneficial over the long term. It is about creating systems and cultures that not only manage knowledge today but also position an organization to continue leveraging its intellectual capital into the future. In a rapidly changing world where knowledge is often the most valuable asset, sustainability in KM ensures that knowledge is not only captured, stored, and shared effectively but also evolves and endures to meet future demands.

Pathways to sustainable and ethical Knowledge management: Knowledge is being learned, shared and applied in order to achieve and maintain competitive advantages and improve individual satisfaction (Li et al., 2021). Sustainable and ethical knowledge management (KM) is essential for organizations seeking to thrive in an increasingly interconnected, resource-constrained, and socially responsible world. To achieve this, KM must not only focus on capturing, sharing, and utilizing knowledge effectively but also consider the long-term implications for people, the planet, and business practices. Here are some pathways to build a sustainable and ethical approach to knowledge management: Uplifting the knowledge sharing collaborations: To foster an organizational culture where knowledge is shared freely and ethically across different levels of management to encourage a culture of openness. To facilitate collaboration in implementing technology with careful attention and to use the same with sense of responsibility. Sharing the knowledge between organizations for developing cross- organizational collaborations and preventing the soiling of valuable insights. Inclusive Knowledge Creation: Creating a diverse perspective that epitomize underrepresented groups and amplify voices for ensuring the collective intelligence of diverse groups. The KM should engage the employees and stakeholders to integrate value that reflects the collective intelligence of diverse groups and not shaped only for narrow interests.

1.7 Aligning KM Practices with Ethical Standards and Organizational Values

To cultivate the belief of ethical practices in academics and industry along with KM, ethical leadership plays a very important role in increasing the organizational performance. It is assumed that ethical leaders are honest and trust worthy (Sosik et al., 2019). Moreover, these people are described as fair as possible, sound decision-makers and as persons who behave ethically in their professional life (Sharma et al., 2019). The important parameters of ethical standards in inculcating KM practices may include morale, values, self-respect, human rights, constructive role etc.

Moreover, the research suggests, that ethical leaders always have inclination towards human centric approach and pay more attention towards the empathic issues and prevent in spreading negativity in the organization. KM practices has been described as playing a very vital role in developing healthy sustainable organizational culture. There are two basic KM processes that work for the betterment and development of ethical standards in the organization: Knowledge System: It is understood that not all the knowledge can be developed within the organization. Some organizations actively get involved in building up positive networking to broaden up the growth and deepen their knowledge domain. The KM practices like knowledge networks can take various forms such as inter and intra group organizational networks, hybrid work culture, virtual teams, onsite experiences etc. The significance of knowledge networks in KM practices is the growth and development of organization with reasonable ethical leaders with prioritizing ethical values like integrity, trust, self-esteem, trustworthiness, will actively promote the utilization of KM practices for development and enhancement of employee performance.

Building a Knowledge Sharing Ecosystem: KM is usually regarded as an individual perspective of sharing knowledge. In organizations, employees are probably afraid of losing their valuable information and insight of their expertise. For creating a healthy knowledge sharing ecosystem, employees in the organization must be educated on the positive and negative aspects of knowledge sharing. The organization can introduce the different initiatives, such as open-door policy, implementing knowledge sharing platforms, actively motivating the employees and rewarding them for exceptional performance, navigating the employees in developing healthy work culture in the organization with career development. To work more efficiently in a positive knowledge sharing ecosystem where employees can support each other and improve their knowledge by exchanging their expertise.

2 Objectives of the Study

- Assesses the Role of Philosophical Knowledge Management (KM) Approaches in Promoting Positive Thinking and Adaptability in Organizations.
- Exploring the Impact of Ethical KM Practices on Employee Engagement and Retention.
- Analyzing the Role of Philosophically fostered Ethical Leadership in Ensuring the Sustainability of KM Practices.

3 Literature Review

Philosophical Underpinnings of Leadership Leadership's foundation can be traced to philosophical theories that emphasize ethics, morality, and practical wisdom. Kant stresses the imperative of duty and universal moral principles in guiding leadership behavior, asserting that ethical leadership stems from adherence to fundamental moral laws. Aristotle, in Metaphysics (n.d.), argues for virtue ethics, advocating that leadership should embody virtues such as courage, justice, and wisdom to inspire ethical conduct among followers. Pragmatism, as advanced by Dewey, J. (1938) and James, W.(1907), highlights the importance of practical experience and adaptability in leadership. **Intersection of Leadership and Knowledge Management**: The interplay between leadership and knowledge management (KM) is pivotal in contemporary organizations. Kulkarni, Ravindran, and Freeze (2006–2007) develop a KM success model, identifying key factors such as organizational alignment, technology integration, and knowledge-sharing behaviors. Szulanski (1996) identifies challenges in knowledge transfer within organizations, describing these as "internal stickiness," which can impede the dissemination of best practices. Ong and Lai (2007) emphasize the critical role of user satisfaction in KM systems, advocating for systems that align with user needs to enhance knowledge-sharing and collaboration. Ethical climates, as discussed by Peterson (2002) and Tseng (2011), significantly influence KM practices. **Cultural and Spiritual Contributions to Leadership:** Leadership inspired by spiritual and cultural philosophies offers unique insights. The teachings of Swami Vivekananda (n.d.), as explored by Chaudhuri (n.d.), present leadership as a selfless pursuit of service, emphasizing resilience, vision, and moral integrity. These principles resonate with Mill (1863) utilitarianism, which advocates for actions that maximize collective well-being. Patagonia's sustainability initiatives (Patagonia, n.d.; Wikipedia Contributors, n.d.) reflect ethical leadership in action, demonstrating a commitment to environmental stewardship and equitable practices. Their approach underscores the integration of ethical values into corporate strategies, setting a benchmark for responsible business leadership. **Broader Implications of Leadership and Knowledge Management**: Sustainability in KM is increasingly vital in a rapidly evolving global context. Moraes (2018) explore how leadership can drive the sustainability of KM practices, ensuring long-term value creation. Dewey (1938) emphasis on experiential learning and Vygotsky (1978) socio-cultural theories further highlight the importance of fostering collaborative environments that encourage continuous learning and development. The role of leadership in crisis management is illustrated in the 9/11 Commission Report (2004), which underscores the importance of adaptability, decisiveness, and learning from past events to enhance future preparedness.

4 Analysis of the Objectives

4.1 When Philosophically Fostered Ethical Leadership Contributes Towards Decision Making and Sustainability in Organizations.

Philosophically driven KM instigates ethical leadership, and that integrates individual's moral principles, their surrounding human atmosphere, philosophically rooted ethics, and values-based decision-making abilities into organizational practices. As mentioned by Brown and Treviño (2006) in their article Ethical leadership: A review and future directions. The Leadership Quarterly, 17(6), 595–616. "Leaders who adopt this approach use framework such as **deontology**, **utilitarianism**, and **virtue ethics** to guide their actions and foster a culture of accountability, fairness, and sustainability". Although it varies on individual basis as every individual has a different journey, ethics, philosophical ideologies and a different outlook towards leading their organizations, some leader follow the concepts of their deontological based decisions, for example- Some leaders might consider company's profits over employee layoffs in the case of some global financial crises or global pandemic, in such cases leaders would consider mass lay off of employees to cut costs, this decision might satisfy that leader's individual ethical standards, but on contrary another individual leader might have a different approach and they might look at this decision as something way too harsh and pressurizing on employees in the time of crises, both financially as well as mentally, such leaders might have a different set of deontological ethics and they might prefer employees wellbeing, job security and happiness over company's short term losses. In both the cases, both the leaders are using their own set of philosophically fostered knowledge and leadership ethics, but there is a strong contrast between both the approaches, this difference in approaches are a direct result of that individual leader's past experiences, journey, philosophical parameters that they've inculcated in them since their childhood, their upbringings, Inclusivity, willingness to prioritize transparency and accountability and inclusive decision making (keeping in mind the shares and contributions of all stakeholders) are again fostered by philosophically fostered ethical values. Such leaders often follow the democratic and transformational based leadership styles. A perfect case example of our discussion is that of **Mr. Ratan Tata,** and his conglomerate Tata Group, Mr. Tata was an famous Indian industrialist, philanthropist, Entrepreneur and former chairman of the **Tata Group,** (one of India's largest and oldest conglomerates) Throughout his career, Mr. Tata has regarded and followed pretty well all the 3 main philosophical parameters deontological ethics, Utilitarianism and virtue ethics, he has always prioritized his duties, what is morally right and ethical decisions that not just sustain his business but also keep his employees satisfied. Tata group has only partnered with suppliers who match the company's ethical standards. And not just that, Tata's commitment to social welfare and wellbeing of his employees throughout his conglomerate is something that can't me missed to discuss about, Ratan Tata ensured that even in tough economic times, Tata Group remained committed to worker rights. For example, the company never engaged in cost-cutting measures that would exploit workers or reduce safety standards, following the duty-based ethics of treating employees with dignity Tata (2017a). If we talk about his Utilitarianist approaches Ratan tata has always ensured that he keeps in mind the interests of all the stakeholders (customers, employees, social communities etc.) For example, TATA Nano car would be a perfect example of that, the prime reasons of launching that car were to ensure that every middle class or middle-income Indian household could fulfill their dream of having their own car, the eco-friendly TATA nano car was launched in Indian back in 2008 at Rupees 100,00. $2000 (As per the 2008 exchange rate of USD vs INR, wherein 1USD equaled 50 rupees). Tata's willingness to watch his country develop is something that's really commendable. Ratan Tata's leadership can also be analyzed through the lens of **virtue ethics**, which focuses on the leader's **character** and **moral virtues**. Tata's personal virtues, such as integrity, courage, humility, and compassion, have shaped the Tata Group's culture Tata (2017b). One of the most notable instances was his handling of the **Tata Corus** and **Jaguar Land Rover acquisitions**. Despite the risks and controversies surrounding these deals, Tata took responsibility for them and emphasized the importance of **long-term sustainability** over short-term financial gains. Khurana (2018). In their journal- Tata Group and its corporate social responsibility practices. Journal of Business Ethics, 150(3), 625–639. Clearly mention that Tata Group has long been recognized for its strong commitment to Corporate Social Responsibility (CSR) and sustainability initiatives, which are deeply embedded in its business practices and corporate philosophy. Under the leadership of Ratan Tata and his successors, Tata Group has taken numerous steps to ensure that its business operations contribute positively to society, the environment, and the communities in which they operate. The group's CSR strategy emphasizes **education**, **healthcare**, **environmental sustainability**, and **community development**. Key initiatives include the work of **Tata Trusts**, which supports health projects like the **Tata Memorial Hospital**, funds educational programs like the **Tata Institute of Social Sciences (TISS)**, and empowers communities through skill development. Tata Group is also committed to **environmental sustainability**. **Tata Steel** focuses on energy efficiency and reducing emissions, while **Tata Motors** promotes **electric vehicles**, exemplified by the **Tata Nexon EV**. **Tata Power** invests in renewable energy projects, including solar and wind power. **Tata Chemicals** advances **green chemistry** to minimize environmental impact in its operations. Tata Group's sustainability goals, the work that they've done is clearly commendable as employees of Tata group are reputedly highly satisfied with the organizational culture, and for any organization employee satisfaction fosters employee's productivity and this again helps the company to sustain its business in the longer run. Tata group's sustainability goals also align with global standards, such as the **United Nations Sustainable Development Goals (SDGs)**, and focus on reducing carbon emissions, increasing renewable energy use, and promoting social welfare. The group is also known for prioritizing **employee welfare**, offering competitive wages and promoting diversity and inclusion. Notable CSR initiatives include the **Tata Sustainability Group**, **Tata's response to COVID-19**, and the **Tata Water Mission**, which focuses on improving access to clean drinking water, especially in rural areas. This is again an example of how Mr. Tata has successfully and ethically used his philosophically fostered knowledge, managed it and used it in a productive, sustainable and socially viable direction.

The Case Example

Another case example can be that of one of the world's most notorious radical Islamic terrorist- Osama bin laden. Hailing from one of Saudi Arabia's richest and most well-known business families, Bin laden himself was a very talented student and civil engineer who held eye catching knowledge about buildings and architecture, but instead of joining his family business or using his knowledge as a civil engineer towards being more productive, and serving for a greater good, Bin laden choose the path of radicalism, violence, and mass destruction. Abdul Azzam played a massive role in turning Bin Laden's ideology towards an extremely radical and violent one. There is a term for this in Hindi which reads as (Sangat Ka Asar) which translates to English as- Company's influence. Bin laden willingly fell for this and used all of his radical philosophies, knowledge of architecture and leadership skills towards planning and execution of the September,11 2002, Terrorist attacks in the United States, that claimed the lives of nearly 3000 people. The only sustainability initiatives that bin laden had in his mind were to promote and sustain his radical religious violent ideologies. The prime reason to choose the case example of Mr. Ratan Tata and Osama bin laden was not to compare and contrast their ideologies, but to prove our point that individuals are leaders and they are different worldwide, places where they are born and raised, the atmosphere, the culture, value system, tolerance, acceptance and other environmental attributes where those individuals are based play a major role in shaping KM, Philosophy, ethical leadership and willingness for sustainability.

5 Discussion

5.1 *Objective 1-Assesses the Role of Philosophical Knowledge Management (KM) Approaches in Promoting Positive Thinking and Adaptability in Organizations*

This objective examines how applying philosophical principles to KM can inspire innovation, creativity, promote a positive mindset towards work related issues, create an overall positive organizational atmosphere and resilience in organizations. It investigates the integration of critical thinking, systems thinking, and mindfulness into KM practices to enhance adaptability and the ability to navigate change. By fostering a culture of continuous learning and open inquiry, having positive mindset, these approaches aim to cultivate environments where employees feel empowered to contribute innovative ideas. Key aspects include exploring how these methods impact individual and collective creativity, support proactive decision-making, and encourage responsiveness to dynamic environments.

5.2 *Objective 2-Exploring the Impact of Ethical KM Practices on Employee Engagement and Retention*

This objective delves into the ethical aspects of KM, focusing on their influence on employee behavior towards their respective tasks, other behavioral aspects related to their colleagues, management people and organizational culture on a whole. It investigates practices such as protecting data privacy, respecting intellectual property, and equitable knowledge sharing between team members, to determine their effects on trust, focusing on being productive at work and not bringing egoistic attitude perspectives, increased job satisfaction, and employee loyalty etc. Ethical KM practices fosters a positive environment where employees feel valued, heard, and secure, which enhances engagement and reduces attrition. Key considerations include assessing how ethical KM practices promote trust, confidence and increase satisfaction. For example- An employee would feel more comfortable productive and satisfied working in an organization that has standardized norms and organizational culture, in other words a well-structured organization, in contrast to if that same employee was to work in a new startup that is still figuring out its foundational stones (the organizational culture, rules, regulations etc.).

5.3 *Objective 3-Analyzing the Role of Philosophically Fostered Ethical Leadership in Ensuring the Sustainability of KM Practices*

This objective talks about how leadership and ethical leadership behaviors fostered by philosophical parameters and styles contribute to the long-term viability of KM initiatives, targeted towards organizational decision making. It focuses on how leaders influence philosophically fostered ethics and KM adoption, usage, and evolution by creating a mission, vision, strong decisive aura, effectively communicating its significance, and addressing the various challenges that arise. Strong leadership fosters alignment between KM objectives and organizational goals, ensuring continuous improvement and relevance. Areas of interest include identifying leadership attitudes that best support KM, strategies for building a shared ethical KM vision, and understanding the barriers leaders face in maintaining ethical and KM systems over time. Baweja, Sudhir Kumar lay their focus on Swami Vivekanand's philosophically empowered ethical Leadership Principles such as empowering & transformational leadership, a clear, inspiring vision: uplifting India through education, spiritual awakening, and self-confidence.

6 Conclusion

Through this study we have tried to explain the integration of philosophical principles into Knowledge Management (KM) practices, highlighting their role in fostering ethical leadership, creativity, and sustainability within organizations. By examining the synergies between philosophy and KM, particularly through ethical considerations and leadership practices, we have discussed that the successful implementation of KM is not solely dependent on operational efficiency but also on the values and ethical frameworks guiding organizational culture. By including the example of Ratan Tata and the Tata Group we have tried to explain and vividly illustrate how a leadership philosophy rooted in ethical principles, such as integrity, social responsibility, and sustainability, can drive long-term success in the business world. Ratan Tata's application of these philosophically grounded KM practices within Tata Industries reflects the potential of combining visionary leadership with a deep commitment to ethical values. By fostering a culture of openness, adaptability, and accountability, focusing on sustainability, Tata has not only ensured the sustainability of his KM practices but also demonstrated that ethical leadership can lead to business success and create positive societal impact. And OfCourse this created a very welcoming organizational culture attracting employes towards working and staying loyal to the organization. Tata's leadership style, characterized by his service-oriented approaches and a focus on the greater good, has enabled Tata Group to navigate complex challenges while maintaining a strong ethical and socially responsible stance. On contrast we also discussed the difference in ideologies and KM management between Ratan Tata and Osama Bin Laden, whilst Ratan Tata's KM practices and leadership fostered growth, trust, and social responsibility, bin Laden's KM practices, its usage and leadership served as an example of how unethical, violent ideologies can destroy societies, destabilize regions, and create a lasting negative legacy. In other words, creating a notion of FEAR. Tata focused on using his philosophically fostered KM practices to create a successful and people centric business empire and sustain his business along with serving people and doing a greater good, Osama bin laden on the other hand used his hard line radical Islamic philosophy and his commendable knowledge of civil engineering in the planning and execution of the September 11 attacks in several ways. His knowledge of building structures helped him understand the vulnerabilities and societal implications of iconic targets like the World Trade Center's Twin Towers for mass destruction in New York City on September 11 2001.

The upshot of the argument is that Philosophy, particularly ontology, plays a foundational role in how knowledge is structured, represented, and managed within organizations. By providing a formal framework for understanding knowledge, ontologies help businesses and institutions better categorize, share, and apply information. The philosophical discussions around existence, categorization, and meaning deeply inform these practices, ensuring that KM systems are both semantically rich and practically useful. Through philosophical engagement, leaders cultivate the virtues and moral reasoning necessary to make principled, just decisions that align with core values—ultimately contributing to the betterment of both their organizations and the wider world. The research has underscored that philosophical foundations in KM can empower organizations to welcome and embrace change and complexity, fostering adaptability while reinforcing ethical considerations that build employee trust and enhance organizational loyalty. Furthermore, leadership that prioritizes ethics and sustainability ensures that KM practices are not just a passing trend but a core element of the organization's long-term strategy. The example of Tata Group demonstrates that integrating philosophical attitudes towards KM often develops ethical leadership, organizations can achieve not only financial success but also meaningful contributions to society, ensuring sustainable growth and long-term relevance in an increasingly dynamic and ethically conscious world. This comprehensive review synthesizes philosophical, theoretical, and practical insights into leadership, emphasizing its ethical and relational dimensions. Leadership's integration with knowledge management highlights its importance in fostering innovation, collaboration, and sustainability. By drawing on diverse perspectives, this review underscores the necessity of ethical leadership in addressing contemporary organizational and societal challenges.

References

Aristotle. (n.d.). *Metaphysics* (Book VII).
Beauchamp, T. L., & Childress, J. F. (2013). *Principles of biomedical ethics* (7th ed.). Oxford University Press.
Brown, M. E., & Treviño, L. K. (2006). Ethical leadership: A review and future directions. *The Leadership Quarterly, 17*(6), 595–616. https://doi.org/10.1016/j.leaqua.2006.10.004
Chaudhuri, A. (n.d.). *Leadership and management of Swami Vivekananda: A world to explore.*
Ciulla, J. B. (Ed.). (n.d.). Leadership ethics: Mapping the territory. In *Ethics, the heart of leadership* (pp. 3–24). Praeger.
Ciulla, J. B., & Forsyth, D. R. (n.d.). Leadership ethics: Mapping the territory. In *The Blackwell companion to business ethics* (pp. 240–257). Wiley-Blackwell.
Dewey, J. (1938). *Experience and education.* Macmillan.
Gadamer, H.-G. (1975). *Truth and method.* Seabury Press.
Greenleaf, R. K. (1977). *Servant leadership: A journey into the nature of legitimate power and greatness.* Paulist Press.
Heidegger, M. (1962a). *Being and time.* Harper & Row.
Heidegger, M. (1962b). *Being and time* (Chapter 1: The problem of being). Harper & Row.
Heifetz, R. A. (1994). *Leadership without easy answers.* Belknap Press of Harvard University Press.
James, W. (1907). *Pragmatism: A new name for some old ways of thinking.* Longmans, Green, and Co.
Kant, I. (1993). *Grounding for the metaphysics of morals* (J. W. Ellington, Trans.). Hackett Publishing Company. (Original work published 1785)
Khurana, A. (2018). Tata Group and its corporate social responsibility practices. *Journal of Business Ethics, 150*(3), 625–639. https://doi.org/10.1007/s10551-018-4027-7
Kulkarni, U. R., Ravindran, S., & Freeze, R. (2006–2007). A knowledge management success model: Theoretical development and empirical validation. *Journal of Management Information Systems, 23*(3), 309–347. https://doi.org/10.2753/MIS0742-1222230309
Mill, J. S. (1863). *Utilitarianism.* Parker, Son, and Bourn.
Michailova, S., & Hutchings, K. (2006). National cultural influences on knowledge sharing: A comparison of China and Russia. *Journal of Management Studies, 43*(3), 383–405. https://doi.org/10.1111/j.1467-6486.2006.00606.x
Moraes, S. de S., Chiappetta Jabbour, C. J., Battistelle, R. A. G., Rodrigues, J. M., Renwick, D. S. W., Foropon, C., & Roubaud, D. (2018). When knowledge management matters: Interplay between green human resources and eco-efficiency in the financial service industry. *Journal of Knowledge Management*, **23**(9), 1691–1707.
Northouse, P. G. (2019). *Leadership: Theory and practice* (8th ed.). Sage Publications.
Patagonia. (n.d.). *Patagonia* [Company website]. Retrieved from https://www.patagonia.com/home/
Saha, S. (2015). Ratan Tata: A study of leadership qualities. *International Journal of Management and Social Sciences Research, 4*(6), 51–56.
Smith, J. (2020). Strategic acquisitions and their impact: A case study of Tata's global ventures. *Journal of Business Strategy, 32*(3), 45–58. https://doi.org/10.1016/j.bushor.2020.03.008
Tata, R. (2017a). *Ratan Tata: The authorized biography.*In: S. K. Gupta (Ed.), Penguin Random House India.
Tata, R. (2017b). Leadership in challenging times. Tata Group Publications.
Unilever. (2023). *Our brands: Doing good through their purpose.* Retrieved from https://www.unilever.com
Vygotsky, L. (1978). *Mind in society: The development of higher psychological processes.* Harvard University Press.
Wikipedia Contributors. (n.d.). *Patagonia, Inc.* Wikipedia. Retrieved from https://en.wikipedia.org/wiki/Patagonia,_Inc
Yukl, G. (2013). *Leadership in organizations* (8th ed.). Pearson.

Quantitative Analysis of the Impact of Loyalty Programs on Consumer Behavior

Ashish Kumar Jha, V. N. Bajpai, and Astha Shukla

Abstract Loyalty programs are strategies used in marketing that benefit consumers in terms of rewards while attracting new customers for companies desiring long term commitment. This paper argues for an investigation into such factors as effects on buyer behavior predetermined by different forms of rewards. Using a variety of businesses in different sectors, the present paper examines how consumer retention, purchase behavior are influenced by different forms of loyalty programs such as tiered rewards, point-based systems as well as member discounts basing their findings on transactional data. The research utilizes an extensive review of data obtained through loyalty program subscribers on an annual basis over a period 12 months. Key factors include how often purchases are made again within set periods, amount spent during each visit/call made within one year and customer life time value (CLV) were taken into consideration so as to determine which among the structures was most efficient in enhancing repetition of transactions. (Jamali, 2008) suggests that customer engagement rates increase significantly when loyalty programs offer customized incentives. Additionally, programs like points, rewards offered at stages show strong linkages with more frequent buys or enhanced payments per individual customer. However, subscription discounts which work well for short periods do not ensure durable customer patronage. The paper provides a smart way of analyzing loyalty programs through the use of data for the purpose of establishing their efficiency in enhancing consumer's commitment and also in order to identify ways of going about this. The findings also elaborate on the broader aspect of loyalty programs'

A. K. Jha
Institute of Technology and Science, Ghaziabad, India
e-mail: ashishkumarjha@its.edu.in

V. N. Bajpai
I.T.S Ghaziabad, Ghaziabad, India
e-mail: vnbajpai@its.edu.in

A. Shukla (✉)
CSJMU, Kanpur, India
e-mail: asthashukladams@gmail.com

V. N. Bajpai
GL Bajaj Institute of Technology & Management, Greater Noida, India

R. S. Mor et al. (eds.), *Sustainable Business Management, Innovation and Technology*, Sustainable Economy and Ecotechnology, https://doi.org/10.1007/978-3-031-98141-8_7

relationship with consumer behavior thus recommending consumer-driven design for tailored programs which should be made for various consumer segments.

Keywords Loyalty programe · Rewards · Customer commitments · Consumer behaviours · Consumer retention · Customer life time value

1 Introduction

In today's cut throat competitive market, organisations are continuously exploring ways to build lasting relationships with customers. Loyalty programs have risen to be a noticeable device in holding back clients and boosting their lengthy attachment to trademarks. These ones are designed to give customers for making purchases repeatedly and remaining faithful, which frequently leads to improve customer pleasure and more profits by companies. Such loyalty programs are expensive to create and maintain thus require an understanding of how they affect buyers for companies spending too much on them.

Qualitative aspects of loyalty programs have been the main focus of many research processes, but their effectiveness needs to be accessed quantitatively. This paper looks at how different types of loyalty programs—points; tiers system for rewarding; membership discounts, etc.-affects customer loyalty and engagement. By using transactional data, the paper tries to understand shopping behavior for those enrolled in loyalty programs and analyzes loops between such programs and various business aspects that are crucial like frequency of purchase and customer value over time.

The primary objective of this research is to determine the effectiveness of loyalty programs in fostering customer retention and enhancing engagement. Additionally, the study explores the role of personalized rewards and the impact of program structure on consumer behavior. By analyzing these aspects, the research aims to provide actionable insights for businesses seeking to optimize their loyalty strategies.

2 Literature Review

The literature on loyalty programs highlights their importance in fostering long-term customer relationships. According to Yi and Jeon (2003), loyalty programs serve as a powerful marketing tool by offering incentives that motivate repeat purchases. These programs are designed to increase consumer switching costs, thereby reducing customer churn. Different types of loyalty programs, including point-based systems, tiered rewards, and subscription-based models, have been shown to significantly impact customer behavior. For instance, Dowling and Uncles (1997) found that point-based systems lead to high purchase frequency, while Verhoef (2003) demonstrated that tiered loyalty programs are more fascinating in driving high-value customer engagement.

More recent studies, such as Kumar and Shah (2014), emphasize the importance of personalized rewards in enhancing customer loyalty. They talked about personalization can enhance program relevance and effectiveness, leading to higher engagement levels. Research by Bijmolt et al. (2011) also supports that loyalty programs are most effective when they offer tangible benefits that align with consumer preferences. Nevertheless, some analysis propose that badly created loyalty schemes can reduce in maximum customer disengagement (Melnyk & Van Herpen, 2013). We are expanding on these previous findings here through quantitatively evaluating how well various types of loyalty programs work.

3 Research Methodology and Hypothesis

This study uses quantitative research designs into unravelling the effects that various kinds of loyalty programs structures have on consumer loyalty and engagement. This research is based on data collected from three major sectors; retailing, hotel industry as well as online business because they depend much on customer loyalty for sustainability and profits. The main objective is to ascertain how different loyalty program schemes including point-based methods, tiered rewards plus membership discounts influence several repeat purchases, transaction value with customer lifetime value (CLV) being an example for key behaviors such as repeat purchase transactions monetary metric over hominine life span years.

3.1 Data Collection

The data has been collected from the transactional records of businesses that have implemented loyalty programs for at least a year. To ensure the reliability of the findings, only customers who have been enrolled in loyalty programs for a minimum of 6 months are included. This allows for the observation of consistent patterns of behavior over time, eliminating biases associated with short-term promotional activities.

The dataset consists of several key variables, including:

Customer demographics: Age, gender, location, and other relevant personal attributes.

Purchase history: The number of transactions, frequency of purchases, and average spend per transaction.

Loyalty program participation details: Type of loyalty program (point-based, tiered rewards, or membership discounts), the level of engagement (e.g., the number of points accumulated), and the nature of rewards received.

This dataset helps for a detailed analysis of how different loyalty program structures influence customer behavior and engagement over time.

3.2 Data Analysis

To measure the impact of loyalty programs on consumer retention and engagement, the study employs several quantitative techniques:

Regression analysis: This method is used to determine the relationships between loyalty program structures and customer retention rates. By controlling for demographic and purchase history variables, regression analysis helps isolate the effects of different loyalty program types on customer behavior.

Analysis of variance (ANOVA): ANOVA is applied to compare the mean values of key metrics, such as purchase frequency, transaction value, and CLV, across the three loyalty program structures. This statistical method helps determine whether differences in customer behavior across loyalty program types are statistically significant.

The study's analytical focus revolves around three key metrics:

Customer retention rates: The proportion of customers who remain engaged with a brand after enrolling in a loyalty program.

Purchase frequency: The number of transactions per customer during the study period.

Average transaction value: The average monetary value of each transaction per customer.

Customer lifetime value (CLV): The total value a customer brings to a business over the entirety of their relationship, influenced by both retention and transaction value.

3.3 Hypothesis Testing

To test the effectiveness of different loyalty program structures, the following hypotheses are formulated:

H1: Loyalty programs with tiered reward structures result in higher customer retention rates compared to point-Based systems and membership discounts.

Tiered reward programs offer escalating benefits, which incentivize long-term engagement. This hypothesis is tested by comparing retention rates across loyalty program structures using regression models and ANOVA.

H2: Personalized rewards in loyalty programs have a positive effect on customer engagement and purchasing frequency.

Personalized rewards cater to individual customer preferences, potentially leading to increased satisfaction and higher purchasing frequency. This hypothesis is evaluated by examining customer engagement and transaction frequency data, particularly among participants in loyalty programs that gives customised rewards.

H3: Point-based loyalty programs lead to increased average transaction value compared to other loyalty program structures.

Point-based systems support customers to spend more to accumulate points, which can later be redeemed for rewards. This hypothesis is tested by analysing the average transaction value among participants in point-based loyalty programs versus those in tiered reward and membership discount programs.

3.4 Objective

The main objective of this study is to provide a comprehensive, data-driven knowledge of how different loyalty program structures influence consumer behavior. By analyzing transactional data using quantitative methods, this research aims to offer actionable insights for businesses to optimize their loyalty strategies, improve customer retention, and increase engagement. The results of this analysis will help organizations better tailor their loyalty programs to the needs and preferences of their customers, thus driving long-term profitability.

4 Data Analysis

This section focuses on analyzing the impact of different loyalty program structures—point-based systems, tiered rewards, and membership discounts—on customer retention, transaction frequency, average spend per transaction, and customer lifetime value (CLV). The data was analyzed by the researchers based on several metrics to determine the performance of individual program types using regression models and ANOVA. This made it feasible to study a variety of consumer behavior alteration mechanisms brought by the loyalty programs in retail, hospitality, and e-commerce industries.

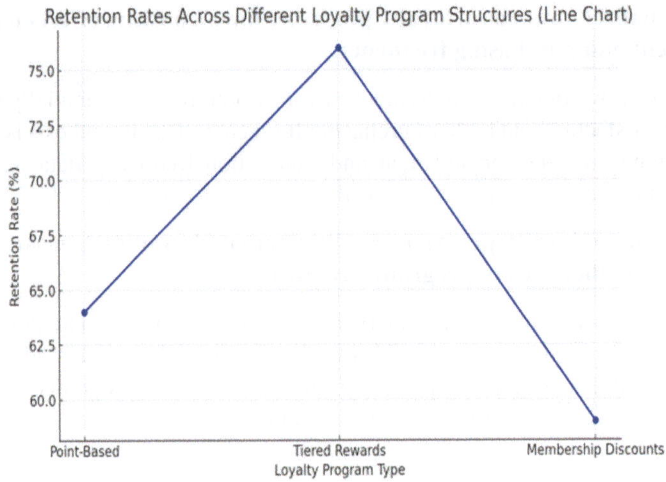

Fig. 1 Retention rates across different loyalty program structures

4.1 Customer Retention

Customer retention is an essential factor that helps to determine whether a customer's loyalty program is successful or not. In loyalty programs, retention is measured as the percentage of customers who still take advantage of an offer even after having joined the program. High amounts of retention therefore imply that clients perceive some benefits associated with brand loyalty schemes hence aspire to establish such lasting connections with a given producer.

In the study, the retention rates for point-based systems, tiered rewards, and membership discounts were compared. The data reveals that tiered rewards programs have the highest retention rate at **76%**, followed by point-based systems at **64%**, and membership discounts at **59%** (Fig. 1).

The significant positive correlation between tiered rewards and retention can be attributed to the escalating benefits that tiered programs offer. As customers move up the tiers, they receive greater rewards, incentivizing them to remain loyal to the brand. The relative success of point-based systems can be linked to their simplicity, allowing customers to accumulate points with each purchase and redeem them for rewards. On the other hand, membership discounts, though beneficial in the short term, do not provide enough long-term engagement to sustain high retention rates.

4.2 Purchase Frequency

Purchase frequency measures how often customers make purchases within a given period. A higher purchase frequency indicates increased engagement with the brand.

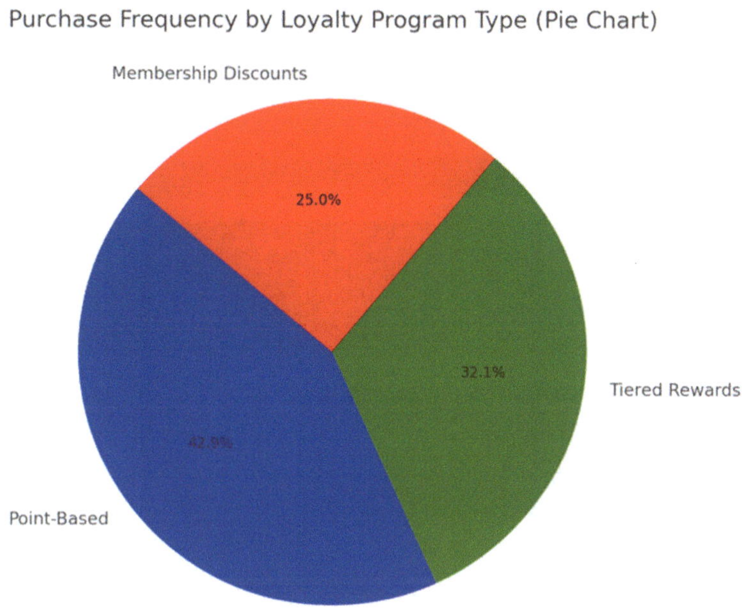

Fig. 2 Purchase frequency by loyalty program type

The results of the interpretation show that **point-based systems** lead to the highest purchase frequency, with customers making an average of **12 transactions per year**, followed by **tiered rewards** with **9 transactions per year**, and **membership discounts** with **7 transactions per year** (Fig. 2).

The higher purchase frequency for point-based systems can be explained by the structure of these programs. Customers are incentivized to make frequent purchases to accumulate points that can be redeemed for rewards. This simple and straightforward approach keeps customers engaged, as they are constantly motivated to make repeat purchases. Tiered rewards programs, although effective in driving higher-value purchases, may not incentivize frequent purchases as effectively as point-based systems. Membership discounts, while attracting initial purchases, do not sustain the same level of engagement, leading to a lower purchase frequency.

5 Average Transaction Value

The average transaction value is another critical metric, measuring the amount of money customers spend per transaction. A higher average transaction value says that customers are spending more on each purchase, adding to increased revenue for the business.

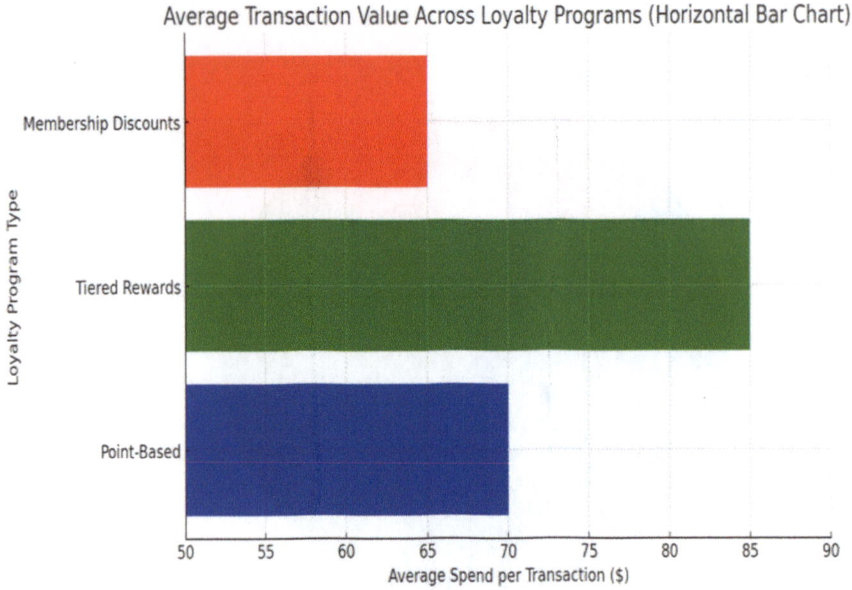

Fig. 3 Average transaction value across loyalty programs

The analysis shows that **tiered reward programs** drive the highest average transaction value, with customers spending an average of **$85 per transaction**, compared to **$70** for point-based systems and **$65** for membership discounts (Fig. 3).

The higher average transaction value for tiered rewards programs can be attributed to the way these programs are structured. To unlock more rewarding rewards, many customers are encouraged by higher levels to buy more items Each transaction will make customers be spurred on by their sense of entitlement as well as continuity in purchases towards spending more amounts on it Point-based rewards also encourage increased expenditure albeit not as much since consumers are able accumulate them every time they buy anything Unlike membership discounts that help consumers save money instantly customer does not enhance what he/she spends per transaction.

6 Customer Lifetime Value

Customer Lifetime Value refers to a key metric through which one can describe how much earnings a business should expect from each customer for the entire time they will remain to be clients to the business (Berger & Nasr, 1998). CLV is calculated using frequency or recency based measures' multiplied by average transaction values and retention rates in absence of personal data (Kinkar, 2006). Assessing overall effectiveness loyalty scheme plan where we include both these issues; that is retention rate within specific period in relation with average transaction

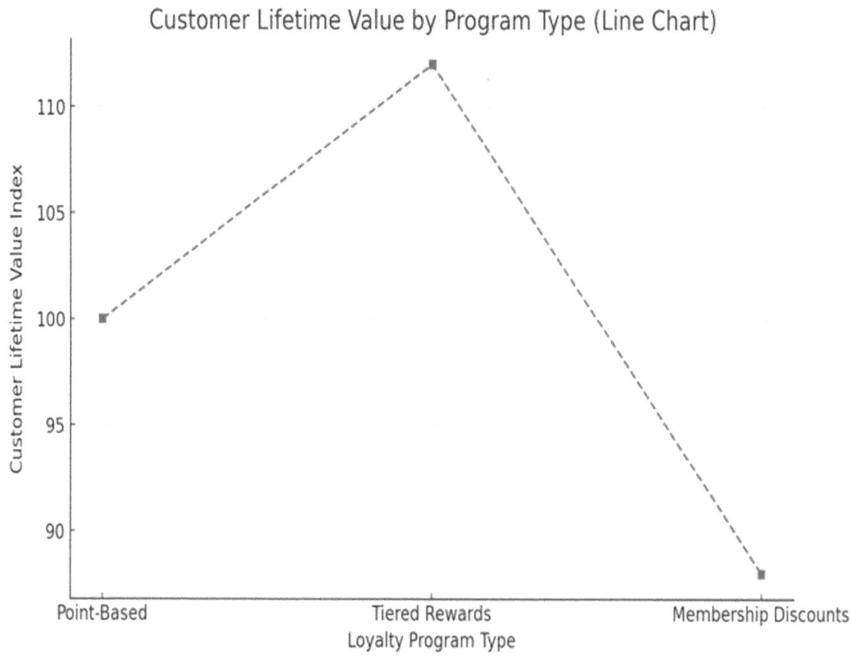

Fig. 4 Customer lifetime value by program type

amount reveals that tiered reward programs generate higher CLVs compared to others with CLVs being 12% higher than other types of program. Also point-based systems as well as membership discounts seem generate lower CLV mainly because they attract lower average transactions which lead to poor retention rates (Fig. 4).

The higher CLV for tiered reward programs can be explained by their ability to drive both higher retention rates and higher average transaction values. As customers move up the tiers, they are to remain loyal to the brand and spend more on each transaction, leading to a higher overall CLV. Point-based systems, while effective in increasing purchase frequency, do not generate the same long-term value due to lower transaction values. Membership discounts, although beneficial in the short term, fail to create the sustained engagement necessary for a high CLV.

7 Data Interpretation

The data clearly demonstrates that different loyalty program structures have varying levels of effectiveness when it comes to influencing consumer behavior. Tiered rewards, for example, excel in driving customer retention and increasing average transaction value, likely due to the appeal of escalating benefits as customers make

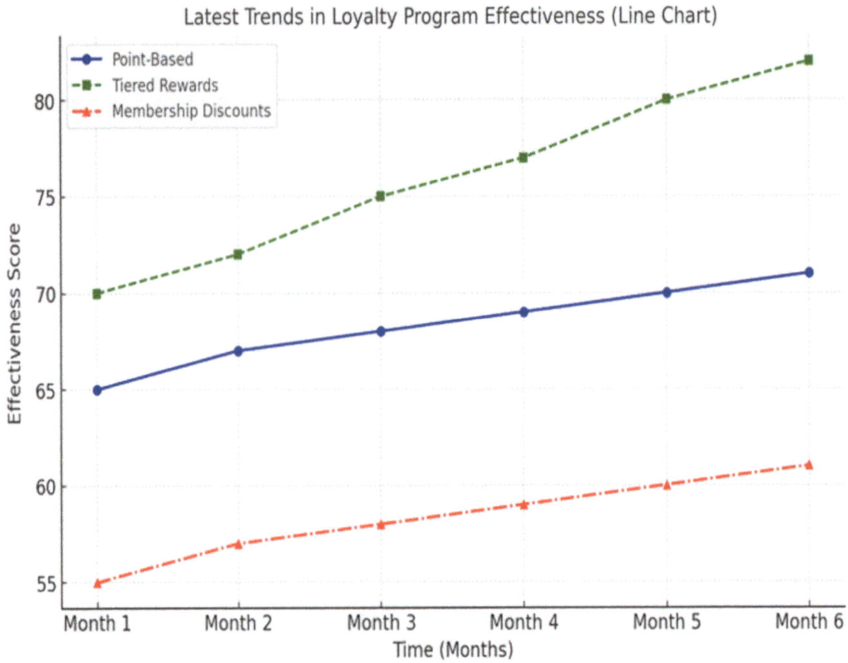

Fig. 5 Latest trends in loyalty program effectiveness

more purchases. This is in line with previous research that suggests tiered systems create a sense of exclusivity, which encourages continued engagement (Fig. 5).

Personalized rewards have a substantial impact on consumer engagement. The data shows that customers receiving tailored rewards based on their preferences has a 22% more engagement rate than those participating in non-personalized programs. This suggests that personalization is a key factor in enhancing the relevance and effectiveness of loyalty programs.

Moreover, the point-based systems, while effective in boosting purchase frequency, do not generate as much long-term value as tiered programs. Although they encourage repeat purchases, the relatively low average transaction value and customer lifetime value suggest that point-based systems may be more suited for short-term promotional campaigns rather than long-term loyalty strategies.

Finally, membership discount programs, while offering immediate savings, do not foster long-term customer loyalty to the same extent as other program structures. The lower retention rates and engagement levels indicate that while discounts may attract initial attention, they are not sufficient to sustain customer loyalty in the long term.

8 Findings

Researchers have recognized several ways in which different types of loyalty programs work differently. The best one of these is multi-tiered programs since they result in a significant spike in client commitment levels coupled with growth within existing transaction amounts and their values too. To encourage clients' continuous participation over a long period at different stages of their development cycle you need rewards that become more and more worthwhile with each succeeding level escalation in the system, which is the most preferable type for such purposes according to customer goals like brand exclusiveness or social recognition, consequently leading to higher commitment levels besides increasing spending propensity.

Although such systems have been effective at increasing the frequency of purchases, they do not perform quite as well when it comes to creating sustainable value over time. They usually make the clients engage in numerous transactions merely to get points. However, the sum total value in terms of transactions is usually lower than those realized using tiered reward systems. This makes them suitable for short-term commitment but ineffective when it comes to building strong and enduring customer relationships.

Personalised rewards: Improving customer satisfaction and engagement significantly depend on giving rewards, which are custom-made. According to research, the 22% more likely customer engagement levels increase when they are personalized because they feel recognized or heard (therefore) resulting into increased customer loyalty and frequency of purchase.

Eventually, **membership discount** schemes appeal to the consumer but do not guarantee loyalty in the long run. The programs stop attracting the clients once they have come in or joined since they do not create an emotional bond strong enough to maintain them thus becoming an ineffective method of driving customer loyalty (save is better).

9 Recommendations and Conclusion

After analysis the obvious conclusion is that businesses wanting to maintain customers over time and increasing purchase amount should utilize the use of steps reward programs as their major loyalty strategies. Such programs make clients stay engaged as rewards increase with activity Progressing in stages through such escalate incentives encourage clients not only to be more active but also spend more money on various purchases thereby greatly aiding in customer retention strategy by companies through regular communication through tier system enables active participation since members are always aware about extra advantages they will acquire as they move up along this line.

This article will suggest reasons why businesses will need to ensure that reward scheme levels have high attraction levels but they are not so difficult for one to achieve; this includes offering personalized incentives at times using other methods besides stages promotions including giving out points.

Stages can be appealing to different customers because it make them feel recognized and help them advance in their lives and so there is need for businesses use this trend in order to retain their clients for sustainable development strategies while others may provide additional discovery experiences too.

References

Bellini, S., Aiolfi, S., & Pellegrini, D. (2021). Loyalty programs and digital rewards: Impacts on consumer engagement. *Journal of Retailing and Consumer Services, 58*, Article 102286.

Breugelmans, E., Bijmolt, T. H., & Zhang, J. (2020). The value of customer loyalty programs in a digital world. *Journal of Marketing, 84*(4), 72–93.

Clark, M., & Melancon, J. P. (2020). The influence of loyalty program rewards on purchase behavior. *Journal of Retailing and Consumer Services, 55*, Article 102115.

Esmark, C. L., Noble, S. M., & Haws, K. L. (2020). Framing loyalty program benefits to enhance perceived value. *Journal of Consumer Psychology, 30*(3), 558–570.

Gai, K., Qiu, M., & Sun, X. (2020). A survey on fintech. *Journal of Network and Computer Applications, 103*, 262–273.

Hwang, J., & Hyun, S. S. (2020). A model of loyalty program effectiveness in the service industry. *International Journal of Hospitality Management, 89*, Article 102608.

Jha, A. K., Bajpai, V. N., Kumar, S., Upadhyay, S.: Impact of social media influencer on consumer decision making. In: *2023 12th International Conference on System Modeling & Advancement in Research Trends (SMART)*, Moradabad, India, 2023, pp. 182–188. https://doi.org/10.1109/SMART59791.2023.10428623

Koo, D. M., & Kim, J. H. (2021). Understanding the effects of mobile loyalty program benefits on customer engagement. *Computers in Human Behavior, 114*, Article 106530.

Liu, X., & Mattila, A. S. (2022). Is customization always effective? The role of customer loyalty program participation. *Journal of Business Research, 140*, 429–440.

Melnyk, V., Van Herpen, E., Trijp, H. C. M., & Tripp, T. M. (2020). The effect of loyalty program introduction on customer purchase behavior. *Journal of Retailing and Consumer Services, 54*, Article 102035.

Pansari, A., & Kumar, V. (2022). The expanding boundaries of marketing: Customization, loyalty, and the power of data. *Journal of Marketing, 86*(4), 1–23.

Yang, S., Lee, K., & Lim, M. (2021). The role of social media in driving customer loyalty in online retail: A study of e-commerce in China. *Electronic Commerce Research and Applications, 47*, Article 101042.

Drivers for Adoption of Industry 5.0 in Food Supply Chain

Simran Malik, Anupama Panghal, Shilpa Sindhu, and Tanvi Rana

Abstract Industry 5.0 represents a transition from the machine-centred approach of Industry 4.0, highlighting the importance of cooperation between humans and advanced technologies to attain sustainable and people-oriented results. This approach strives to incorporate human elements back into the industrial ecosystem, promoting sustainable development, social acceptability, and environmental oversight. Advanced technologies such as human-centric AI, Green IoT, sustainable blockchain, cyber-physical systems, digital twins, and smart sensors have the potential to support efficient food supply chain management. These innovations can enhance resilience, sustainability, and performance while keeping the costs low. This study conducts a comprehensive analysis of the factors acting as motivators for implementing Industry 5.0 technologies by companies. The drivers were assessed and prioritised using the Fuzzy-Analytic Hierarchy Process (FAHP). The study highlights the importance of supply chain resilience as the top priority, followed by need for sustainability and cost savings. This research emphasizes the significant impact of Industry 5.0 on developing robust and eco-friendly food supply chains, highlighting the crucial role of integrating human-centred technology.

Keywords Industry 4.0 · Industry 5.0 · Fuzzy-analytic hierarchy process (FAHP) · Internet of everything (IOE) · Food supply chain

1 Introduction

Industries are often associated with the technologies employed by them, and each technological change is associated with a specific industrial revolution (Hoosain et al., 2020). The first technological change was associated with mechanization and

S. Malik · A. Panghal
National Institute of Food Technology Entrepreneurship and Management, Kundli, India

S. Sindhu (✉) · T. Rana
Maharaja Surajmal Institute (Affiliated to Guru Gobind Singh Indraprastha University), Janakpuri, New Delhi, India
e-mail: shilpasindhu@msijanakpuri.com

R. S. Mor et al. (eds.), *Sustainable Business Management, Innovation and Technology*, Sustainable Economy and Ecotechnology, https://doi.org/10.1007/978-3-031-98141-8_8

steam power to empower industries, known as Industry 1.0. The second technological change was associated with assembly lines, mass production, and electricity, known as Industry 2.0. The third technological change was associated with automation with the help of computers and electronics, known as Industry 3.0. The fourth, at present, is associated with digitalization and is considered the age of artificial intelligence (AI), the Internet of Things (IoT), and cybersystems (Rüßmann et al., 2015). It also helps build strong future leaders, by identifying the skills and competencies associated with digital technology leaders (Fonseca et al., 2021).Industry 4.0 is associated with rapid digitalization, increased unemployment, and dangerous layoffs. Therefore to bring humans again to the forefront in industries, the transformation from industry 4.0 to 5.0 occurred (Mekkunnel, 2019).Industry 5.0 focuses on the concept that robots are not the competitors of humans (Nahavandi, 2019). The role of robots as collaborators of humans should be focused on the fact that competition and challenges should be fought together using human-centric AI, not against each other (Doyle-Kent & Kopacek, 2021). The aim of Industry 5.0 is sustainable development, social acceptability, and environment management and oversight using humans, unlike Industry 4.0, wherein smart machine manufacturing and optimizing machines were considered (Agrawal et al., 2023). In Industry 4.0, human interaction was kept to a bare minimum, and therefore, issues in supply chain management, data privacy, and integration of human resources and machines arose (Khan et al., 2023). Industry 5.0 has an additional vision, which is a value addition to the associated industries and supply chain, along with a human-centric vision, making it undefeatable even beyond the integration of the circular economy and Industry 4.0 smart technologies (Atif, 2023). The technologies associated with Industry 5.0 include the Internet of Things (IoT) built on a broad wireless network known as Green IoT; big data analytics, which uses AI and cobots; smart sensors for increasing the database of big data, and AI-based cobots controlled by humans (Özdemir & Hekim, 2018). Industry 5.0 technologies also include digital twins and sustainable blockchain (Lv, 2023; Wang et al., 2023). Industry 5.0 focuses on the concept of making humans one of the core competencies of the universe. It is increasing human capabilities using artificial intelligence (AI) (Skobelev et al., 2021). Therefore, artificial intelligence and smart machines will be required to increase the efficiency and effectiveness of humans. A similar interaction of Industry 5.0 is also observed in the supply chain to face adverse climate, war, or pandemic situations (Agrawal et al., 2023). The need for Industry 5.0 was realised during pandemic, wherein completely automated machines were not proficient enough to handle food supply chains, and the need for predictive intelligence was felt. Predictive intelligence was possible only through the human-centric approach of sustainable growth and development (Madhavan et al., 2022).

Addressing the global food security problem is one of the significant roles of Industry 5.0, which can be resolved by integrating IoT and Blockchain (Xiao et al., 2023). Another role can be simplicity and quick data accessibility throughout the food supply chain. It makes upstream data of different food players available and facilitates studying the life cycle assessment of different food industries (Rahimifard et al.,

2022). Industry 5.0 can play a vital role in eliminating hunger and improving malnutrition conditions where resources are limited and the population is proliferating. It can also protect against environmental risks and changing climates (Baryshnikova et al., 2022).Lastly, the role of Industry 5.0 technologies also expands to include sustainable agriculture production, renewable resources, bionics, robots, and human collaboration to make cobots that can be coordinated with humans and robots (Demir 2019). Other technologies currently being developed and part of Industry 5.0 include shopfloor trackers, virtual training, advanced sensors, and machine cognition. Their full potential is yet to be harnessed (Paschek et al., 2019).

This paper aims to explore different technologies from Industry 5.0 that can be adopted in food supply chain management. It also reviews the existing literature to discover the reasons why Industry 5.0 is needed in food supply chain management. The drivers for adopting Industry 5.0 were identified through a systematic literature review. Drivers were ranked and prioritized through Fuzzy-AHP, a multiple-criterion decision-making technique. This study majorly focused on followed objectives (a) *To review the technologies associated with Industry 5.0 in food supply chain; (b) To rank and prioritize different drivers for adoption of Industry 5.0 technologies in food supply chain.*

2 Literature Review

Understanding Industry 5.0 is incomplete without the thorough knowledge of technology associated with it. Industry 5.0 concept entails that all technologies should be monitored and controlled by humans (Nahavandi, 2019). As Industries transform from one industrial revolution to another revolution, the digital technologies also undergoes similar transformation. AI of Industry 4.0 transformed to Human-centric AI in Industry 5.0 based on one of the pillars of Industry 5.0 which is human-centricity (Hangl et al., 2022; Wang, et al., 2022). IoT of Industry 4.0 changed to Green IoT in Industry 5.0 which is based on another pillar of Industry 5.0 that is sustainability (Albreem et al., 2021; Arshad et al., 2017; Fraga-Lamas et al., 2021; Zhu et al., 2015). Similarly blockchain of Industry 4.0 changed to sustainable blockchain in Industry 5.0 (Song et al., 2019; Wang et al., 2023). The third pillar of Industry 5.0 is resilience which can only be prevailed by trust and integrity among the producers and consumers. So, in order to make digital technologies much more resilient, cyber-physical systems (CPS) were developed (Lisitsyn et al., 2024). CPS ensured data security and prevent any security breach so that overall efficiency and effectiveness can be increased. With augmented reality (AR) and virtual reality (VR), digital twins (Rahmani et al., 2024) also arose along with improved sensors so that more advanced technologies can be used. In Industry 5.0, all technologies are conformed to the three pillars pertaining to human-centricity, sustainability and resilience. These technologies are summarised as follows:

2.1 Technologies Associated with Industry 5.0

2.1.1 Human-Centric AI

Artificial intelligence (AI) is how humans make machines work based on existing data. Humans make mistakes; therefore, while using AI, a pattern of errors can be observed in artificially processed data. To remove the pattern of errors, a professional must monitor all the mistakes. When we use a human-centric approach for AI, it not only becomes more reliable but also sustainably resilient Wang, et al., 2022). The reasons for adopting AI in the supply chain are cost reduction, improved efficiency, and time and resource reduction. In contrast, the challenges for AI adoption are supply chain resilience, supply chain performance, and AI compatibility (Hangl et al., 2022).

2.1.2 Green IoT

The Green IoT (G-IoT) refers to the implementation of energy-efficient procedures in both hardware and software of IoT technologies (Zhu et al., 2015). The procedures outlined here focus on reduction of the greenhouse effect caused by current applications and services and addressing the impact of the greenhouse effect within the IoT ecosystem (Albreem et al., 2021). One possible application of IoT technologies is reduction of the greenhouse effect, while another focus is on maximizing the environmental efficiency of IoT greenhouses (Arshad et al., 2017). Furthermore, it is crucial for the life cycle of a G-IoT system to prioritize environmentally conscious design, production, utilization, and disposal/recycling to minimize its environmental impact (Fraga-Lamas et al., 2021).

2.1.3 Sustainable Blockchain

Blockchain has numerous advantages in food supply chain management, such as better future forecasting, more transparency and traceability, and thorough auditing of the goods flown within the various supply chain steps (Song et al., 2019). Blockchain technology, known for its immutability, transparency, and traceability, is being increasingly adopted by enterprises to improve sustainability practices. Its potential impact on supply chain management is significant, with researchers focusing on integrating blockchain technology into supply chain systems. Advancements in artificial intelligence and the Internet of Things have enabled continuous monitoring and intelligent management of supply chain systems, which led to the formation of sustainable blockchain (Wang et al., 2023).

2.1.4 Cyber-Physical Systems

Cyber-physical systems (CPSs) are intricate systems which combine physical, biological, and computational components through engineering networks. They can improve information flow between industrial equipment, enhance chains of production, and business management operations and customer relationships. CPSs can monitor and control the entire production process, ensuring production aligns with customer preferences. Key technology driven trends of CPSs include artificial intelligence (AI), machine learning (ML), big data analytics, augmented reality, the Internet of Things (IoT), quantum computing, fog computing, 3D printing, modelling, and RFID tags. Implementing CPSs can significantly improve production efficiency and quality by collecting data on raw materials, temperature, technological conditions, and food product readiness (Lisitsyn et al., 2024).

2.1.5 Digital Twins

Digital twins portray a crucial role in facilitating the real-time modelling and simulation of physical systems. An in-depth examination of the digital transformation and digital twin concepts, emphasizing their significant impact on both local and global Future of Manufacturing (FoF) initiatives. Automated processes are essential for enhancing productivity, improving production control, and reducing challenges for workers. These factors collectively contribute to more efficient process management (Rahmani et al., 2024).

2.1.6 Smart Sensors

Industry 5.0 emphasizes cutting-edge smart sensors, such as optoelectronic and mechatronic sensors, and processes driven by artificial intelligence. It presents a comprehensive model for an Industry 5.0 framework, which includes the integration of technological advancements and hypothetical manufacturing lines. The smart sensor model encompasses the simulation of information flow, the development of business process modelling and notification workflows, and a comprehensive BPMN model for robotic and machine control, predictive maintenance, and quality assessment (Massaro, 2023).

Knowing the technologies associated with Industry 5.0, the need to study what motivates adoption of advanced technologies associated with latest industrial revolutions arose in food supply chain.

2.2 *Drivers for Adoption of Industry 5.0 in Food Supply Chain Management*

Industry 5.0 follows a triple bottom-line approach to achieve sustainability in supply chains (Varriale et al., 2023), through which people's corporate social responsibility (CSR) towards the food supply chain has been recognized. There has been a growing focus on managing supply chains in an environmentally friendly way and fostering sustainable innovations (Hu & Chen, 2023). The literature review highlights various factors related to Industry 4.0, AI, IoT, green supply chain, digital technologies, blockchain, digital twins, smart sensors, big data analytics, and sustainable supply chain management. The prominent drivers for the adoption of Industry 5.0 in the food supply chain highlighted in literature are discussed as follows:

1. *Supply chain resilience (D1):* Systems which prioritize the ability to handle unexpected events or disruptions should be given more consideration. For example, it is essential to prioritize e-traceability systems that allow for prompt responses to food safety incidents or disruptions in the supply chain (Wang, 2022; Singh, 2023).
2. *Sustainability (D2):* Sustainability is crucial in safeguarding the agricultural climate and promoting environmentally friendly practices in the food supply chain. These practices positively impact the economy by reducing recalls, enhancing social welfare, and mitigating food waste. Ensuring environmental protection is a crucial prerequisite for obtaining the sustainability certificate (Faisal, 2016; Wang, 2022; Krstic, 2023; Singh, 2023).
3. *Corporate Social Responsibility (CSR) (D3):* CSR is a motivating force that encompasses social responsibility and ethical values. It is impossible to exaggerate the significance of companies investing in moral and social responsibility throughout their industrial processes. It is possible to promote equitable labour practices, mitigate environmental damage, and foster diversity and inclusion in the workplace (Flammer, 2013; Eccles 2014; Zhang, 2023).
4. *Supply chain performance (D4):* The driver focuses on optimizing operations between different agents in the food chain and improving process management to ensure a return on investment (Wang, 2022; Singh, 2023).
5. *Cost Saving (D5):* Industry 5.0 seeks to reduce operational costs by addressing issues pertaining to communication networks, waste management, supply chains, industrial process optimization, and other associated obstacles. Minimizing expenditures is beneficial for cost savings, which may serve as a driver for the adoption of Industry 5.0 (Gutierrez et al., 2015; Zhang, 2023).

The literature review summarises the drivers for the adoption of Industry 5.0 in the food supply chain to identify the main motivating factors of transformation towards Industry 5.0. The next section examines the drivers to rank and prioritise them. The methodology provides a framework for drivers based on their relative importance in the food supply chain which is explained in the section given below.

2.3 Methodology

The methodology adopted in the study can be discussed as a five-step method. The first step involves a thorough analysis of existing literature for determining the drivers required for the adoption of Industry 5.0 in the food supply chain. The next step is to identify the experts in the domain of the food supply chain who can validate and analyse the identified drivers of Industry 5.0 adoption. For this study, five experts were approached for the said purpose. These five experts included three academicians and two industry representatives. These experts were provided with a structured questionnaire to get their responses. In the first phase, a pairwise comparison of the five drivers required for adopting Industry 5.0 in the food supply chain was carried out. The formula for determining the number of comparisons is n(n-1)/2. Then a matrix was prepared, which is generated by assessing the significance of the alternatives according to each criterion, utilizing a nine-point scale (1–9). The scale ranges from 1/9, indicating a lower value, to 1, representing equal importance, and up to 9, signifying the most importance. 3 represents more importance, 5 represents high importance and 7 represents extreme importance. 2, 4, 6 and 8 represent importance in between their scale ranges. The third step involves computing the criteria weightage of the drivers for the adoption of Industry 5.0 based on the AHP method. The step analyses the AHP matrices of experts. The next step is determining the fuzzy scale. In this case, a 9-point fuzzy scale is used to convert it into linguistic terms and for each pairwise comparison, the lowest value, medium value and highest value are calculated. Then based on the fuzzy scale, the criteria weightage is computed by the fuzzy AHP method. The last step depicts the comparison of criteria weightage for all the drivers based on the AHP and fuzzy AHP method. The weightage are almost similar but fuzzy AHP is preferred over AHP because of more reliable and consistent values. Then, all the drivers are ranked based on AHP and fuzzy AHP for the adoption of Industry 5.0 in the food supply chain. The process of decision-making is aimed at selecting the best alternative out of all the given alternatives based on AHP and fuzzy AHP method (Pomerol & Barba-Romero, 2000).

Analytic Hierarchy Process (AHP)

MCDM techniques are mostly employed to tackle complex problems that necessitate evaluating and choosing suitable alternatives based on their performance across multiple predefined criteria. The Analytic Hierarchy Process (AHP) has gained significant traction and is now widely recognized as a more reliable approach in its subsequent iterations. Utilizing the hierarchy approach provides numerous advantages that enable the incorporation of various criteria through consensus, facilitate the logical integration of data and human judgment into the model, and simultaneously resolve complex problems involving multiple criteria and alternatives. In addition, it aids in reducing bias within the system by offering a systematic method for quantifying both tangible and intangible data. AHP has been recognized as a valuable tool within the MCDM process. It is appreciated for its user-friendly interface and its capability to convert subjective feelings into prioritized alternatives (Avikal,

Mishra, and Jain 2013; Sopha et al., 2016). The matrix for the drivers required for the adoption of Industry 5.0 in the food supply chain is given below (Table 1):

After the collection of matrix responses, data is used for pairwise comparison. The matrix is normalised in the first step along both the rows and columns. The number of steps it takes for normalisation helps determine the eigenvalue, which provides the decision matrix.

Decision Matrix: The primary eigenvector of the decision matrix is of utmost importance in the AHP for establishing priority weights. The AHP employs pairwise comparisons in order to construct a decision matrix. The eigenvector is derived from this matrix in order to ascertain the relative significance of various criteria and alternatives. The method mentioned above, despite being widely used, has faced criticism for its deterministic nature, which neglects the potential for errors in expert evaluations (Mohammed, 2023). Recent developments have been made to enhance the creation of the pairwise comparison matrix in the Analytic Hierarchy Process (AHP). The objective of these modifications is to reduce the degree of subjectivity associated with the assignment of weights. Moreover, the integration of fuzzy logic into the AHP, as demonstrated in multiple applications such as the evaluation of tourism trends, highlights the versatility of this approach in efficiently handling uncertainty and improving the precision of rankings (Saifullah, 2019). The principal eigenvector is of significant importance in the determination of weights within the Analytic Hierarchy Process (AHP) as a whole. Nevertheless, ongoing research endeavours are focused on enhancing the process by addressing its inherent limitations. The resulting weights for the criteria based on the pairwise comparisons for AHP are calculated for each expert. The average of all priority values of each driver is taken, and the final average of all tables for AHP is given below in the table and rank is placed by taking into consideration the AHP priority percentage. The one with the highest priority percentage will be ranked 1, second most ranked two and similarly, for all the drivers, the rank is calculated (Table 2).

Fuzzy-Analytic Hierarchy Process (FAHP)

On the other hand, Fuzzy-AHP integrates fuzzy logic as a means to handle the inherent vagueness and imprecision in human decision-making. This enables a more detailed depiction of preferences through the use of fuzzy sets and membership functions (Poongothai et al., 2023). The integration of Fuzzy-AHP allows for more

Table 1 Direct relation matrix

Drivers	D1	D2	D3	D4	D5
D1	1	n_1	n_2	n_3	n_4
D2	$1/n_1$	1	n_5	n_6	n_7
D3	$1/n_2$	$1/n_5$	1	n_8	n_9
D4	$1/n_3$	$1/n_6$	$1/n_8$	1	n_{10}
D5	$1/n_4$	$1/n_7$	$1/n_9$	$1/n_{10}$	1

Table 2 AHP criteria weightage and priority percentage

Notation	Drivers for adoption of Industry 5.0	Weightage	AHP priority percentage (%)	AHP rank
D1	Supply chain resilience	0.4582	48.52	1
D2	Sustainability	0.2656	26.56	2
D3	Corporate Social Responsibility (CSR)	0.1398	13.98	3
D4	Supply chain performance	0.0646	6.46	4
D5	Cost Saving	0.0446	4.46	5

effective incorporation of the intricacies associated with human factors and team dynamics, especially in domains such as software development, where uncertainty is widespread. According to research findings, Fuzzy-AHP has been shown to achieve slightly higher levels of accuracy in scenarios of real-world in comparison to traditional AHP. That can be attributed to its ability to overcome the limitations of expert judgment by incorporating a wider range of decision-making inputs (Wongvilaisakul et al., 2023). Therefore, although the Analytic Hierarchy Process (AHP) offers a systematic methodology, Fuzzy-AHP improves adaptability and authenticity in decision-making procedures. A 9-point fuzzy crisp scale is used to convert it into linguistic terms, and for each pairwise comparison, the lowest value, the medium value, and the highest value are calculated. Then, a 5*5 matrix is converted into a 5*15 matrix where, for each cell, there are three different values. Fuzzy AHP is used because it is much more efficient and more reliable than normal AHP. However, the values are almost similar (Bakır & Atalık, 2021). The crisp scale for a fuzzy triangular matrix is given below (Table 3):

The calculation for fuzzy AHP is as follows:

Step 1: Convert the original matrix to a Triangular fuzzy scale matrix. It is replacing 1, 1, 1 as 1, 1, 1 and similarly for all the triangular fuzzy scale values of the given

Table 3 Fuzzy Triangular matrix and its scale

Crisp scale	Triangular fuzzy scale	Reciprocals
1	(1,1,1)	(1,1,1)
2	(1,2,3)	(0.33,0.50,1.0)
3	(2,3,4)	(0.25,0.33,0.50)
4	(3,4,5)	(0.20,0.25, 0.33)
5	(4,5,6)	(0.17, 0.20, 0.25)
6	(5,6,7)	(0.14,0.17,0.20)
7	(6,7,8)	(0.13,0.14,0.17)
8	(7,8,9)	(0.11,0.13,0.14)
9	(8,9,9)	(0.11,0.11,0.13)

table of the responses. Thus, the normal matrix response is converted to a triangular fuzzy scale matrix.

Step 2: Find the Geometric mean (GM) for the resultant triangular fuzzy scale matrix with the given formula:

$$GM = \{(x1)(x2)(x3)...............(xn)\}1/n$$

where, x = individual paired weight value of individual expert, n = Sample size (number of judgments (Mondal et al., 2023).

Step 3: Then, find the weightage for the given GM. Because the weightage is less than or more than 1, it undergoes normalization where all individual weightages of all drivers are divided by sum total of all the weightages so that the sum total of all the weights is equal to 1.

The weightage of the given vector is given by the equation

$$W = \left(d'(S1), d'(S2), \ldots, \ldots, .., d'(Sn)\right)\tau$$

Step 4: Normalize the weights and give fuzzy ranks for the following weightage similarly for all the expert opinions and then the average of all the fuzzy responses is taken which is depicted in the table given below:

The normalized weight vector W can be obtained by the given below equation

$$W = (d(S1), d(S2), ---, d(Sn))T$$

where W signifies a non-fuzzy number calculated for each comparison matrix.

The fuzzy criteria weightage computed after the following steps is as given in Table 4 below:

The Fuzzy Analytic Hierarchy Process (FAHP), on the other hand, incorporates the use of fuzzy logic to efficiently handle these uncertainties, thereby facilitating a more advanced evaluation of the crucial determinants of success in the fields of digital marketing and supply chain resilience (Wiwatkajornsak & Phuaksaman, 2024). The capacity to adjust is of paramount significance for food industries that face difficulties in adopting Industry 4.0 and Industry 5.0 technologies, as indicated by the organizational and strategic challenges observed in the Indian context (Lahane et al., 2023). Thus, the FAHP method presents itself as a more effective instrument for assessing drivers for the implementation of Industry 5.0, thereby improving decision-making in intricate situations (Fuentes et al., 2024).

The present study investigates the application of AHP and FAHP in the context of Industry 5.0 adoption within the food supply chain. The findings highlight the distinct advantages of FAHP in efficiently handling the inherent uncertainties associated with the process of decision-making. The conventional AHP provides a structured framework for prioritizing the factors that impact the adoption of Industry 5.0.

Table 4 Fuzzy AHP Rank and normalized weightage determination

Drivers for adoption of Industry 5.0	Low-weightage	Medium-weightage	High-weightage	Weightage	Normalized weightage	Fuzzy rank
Supply chain resilience	0.259281	0.434338	0.672778	0.455466	0.421019053	1
Sustainability	0.196104	0.313031	0.527493	0.345543	0.319409395	2
Corporate Social Responsibility (CSR)	0.085531	0.141192	0.243203	0.156642	0.144779508	3
Supply chain performance	0.042102	0.068529	0.114802	0.075145	0.069461395	4
Cost Saving	0.029302	0.04291	0.074856	0.049023	0.045315077	5

Table 5 Comparison of AHP and Fuzzy AHP

Drivers for adoption of I5.0	AHP priority (%)	AHP rank	Fuzzy AHP weightage %	Fuzzy AHP rank
Supply chain resilience	48.52	1	46.94	1
Sustainability	26.56	2	27.64	2
Corporate Social Responsibility (CSR)	13.98	3	14.17	3
Supply chain performance	6.46	4	6.65	4
Cost Saving	4.46	5	4.60	5

Nevertheless, it is of paramount importance to note that the aforementioned statement may not comprehensively capture the complexities and uncertainties that are inherent in expert assessments within dynamic environments, such as the food supply chain, which has undergone significant disruptions due to the COVID-19 epidemic and other related factors (Hsu et al., 2024; Mandal et al., 2023).

3 Result and Discussions

Supply chain resilience is one of the major drivers for adoption of Industry 5.0 in the food supply chain, with almost 50% of the total weightage (Hsu et al., 2024). It has priority over the rest of the other drivers. After that, sustainability is the second major driver required for the adoption of Industry 5.0 (Vasanthraj et al., 2023). The result shows that cost saving is the least important driver and is at the bottom when considered one of the less influential reason for adoption of Industry 5.0 in the food supply chain (Nahavandi, 2019).Corporate Social Responsibility (CSR) and Supply chain performance are the mediocre driving force behind the adoption of Industry 5.0 in the food supply chain with the weightage of 14.17% and 6.5% in Fuzzy AHP and 13.98% and 6.46% in AHP with the third and fourth rank respectively. The study discusses that social responsibility and ethical values can only drive businesses to a limited extent to invest in Industry 5.0. Similarly, optimisation of the food chain measured by their performance also has limited influence on the food supply chain for adoption of Industry 5.0 (Table 5).

4 Conclusion

The study reveals a comprehensive explanation of the drivers for the adoption of Industry 5.0 technologies in food supply chain and management. The importance of supply chain via sustainability and resilience in the current and future industrial context is demonstrated by the prioritization of these factors through the utilization of the AHP and FAHP methodologies. The integration of Industry 5.0 technology will be crucial in ensuring the resilience, sustainability, and efficiency of the evolving of food supply chain. Majorly, supply chain resilience is one of the prominent drivers required for the adoption of Industry 5.0 in the domain of food supply chain. If businesses want to increase their efficiency, their main focus should be on supply chain resilience followed by sustainability. The study findings revealed that Industry 5.0 adoption in the food supply chain is almost 75% dependent on supply chain resilience and sustainability. Therefore, more comprehensive studies on supply chain resilience and sustainability are required for the implementation of Industry 5.0 in the food supply chain. Despite these numerous advantages, there are a few disadvantages like cost saving which was thought to be the major driver for implementation of Industry 5.0 comes out to be the least favourable driver out of the five major drivers in the given study. The traditional methods are still favoured over the modern methods and it is despite the sustained advantage of sustainability in the modern methods.

5 Future Research Scope

Future research scholars and industry experts should go beyond their scope and expand their areas of research to broader horizons. The studies should also focus on longitudinal and horizontal areas of research. Managers and other academic professionals can use the given research for determining the major drivers required for the adoption of Industry 5.0 in the food supply chain. The research is also useful for decision-makers and policymakers in implementing Industry 5.0 in the food supply chain. The research can be used for further case studies and determining the challenges of Industry 5.0. Future studies can also include the study of demographic variables along with the major drivers. Such studies can motivate other scholars to also contribute to similar areas of research.

References

Albreem, M. A., Sheikh, A. M., Alsharif, M. H., Jusoh, M., & Yasin, M. N. M. (2021). Green Internet of Things (GIOT): Applications, practices, awareness, and challenges. *IEEE Access, 9*, 38833–38858. https://doi.org/10.1109/access.2021.3061697

Agrawal, S., Agrawal, R., Kumar, A. et al. (2023) Can industry 5.0 technologies overcome supply chain disruptions?—a perspective study on pandemics, war, and climate change issues. *Operations Management Research*. https://doi.org/10.1007/s12063-023-00410-y

Arshad, R., Zahoor, S., Shah, M. A., Wahid, A., & Yu, H. (2017). Green IoT: An investigation on energy saving practices for 2020 and beyond. *IEEE Access, 5*, 15667–15681. https://doi.org/10.1109/access.2017.2686092

Avikal, S., Mishra, P., & Jain, R. (2013). A Fuzzy AHP and PROMETHEE method-based heuristic for disassembly line balancing problems. *International Journal of Production Research, 52*(5), 1306–1317. https://doi.org/10.1080/00207543.2013.831999

Atif, S. (2023). Analysing the alignment between circular economy and industry 4.0 nexus with industry 5.0 era: An integrative systematic literature review. *Sustainable Development, 31*(4), 2155–2175. https://doi.org/10.1002/sd.2542

Bakır, M., & Atalık, Ö. (2021). Application of Fuzzy AHP and Fuzzy MARCOS approach for the evaluation of E-Service quality in the airline industry. Decision Making. Applications in Management and Engineering/Decision Making: *Applications in Management and Engineering, 4*(1), 127–152. https://doi.org/10.31181/dmame2104127b

Baryshnikova, N., Altukhov, P. L., Найденова, Н. В., & Shkryabina, A. (2022). Ensuring Global Food Security: Transforming Approaches in the Context of Agriculture 5.0. *IOP Conference Series: Earth and Environmental Science, 988*(3), 032024. https://doi.org/10.1088/1755-1315/988/3/032024

Demir, K., Doven, G., & Sezen, B. (2019). Industry 5.0 and human-robot co-working. *Procedia Computer Science, 158,* 688–695. https://doi.org/10.1016/j.procs.2019.09.104

Doyle-Kent, M., & Kopacek, P. (2021). Adoption of collaborative robotics in industry 5.0. An Irish industry case study. *IFAC-PapersOnLine, 54*(13), 413–418. https://doi.org/10.1016/j.ifacol.2021.10.483

Eccles, R. G., Ioannou, I., amp; Serafeim, G. (2014). The impact of corporate sustainability on organizational processes and performance. *Management Science, 60*(11), 2835–2857. http://www.jstor.org/stable/24550546

Faisal, M. N., & Talib, F. (2016). Implementing traceability in Indian food-supply chains: An interpretive structural modeling approach. *Journal of Foodservice Business Research, 19*(2), 171–196. https://doi.org/10.1080/15378020.2016.1159894

Flammer, C. (2013). Corporate social responsibility and shareholder reaction: The environmental awareness of investors. *Academy of Management Journal, 56*(3), 758–781. https://doi.org/10.5465/amj.2011.0744

Fonseca, L., Amaral, A., & Oliveira, J. (2021). Quality 4.0: The EFQM 2020 model and industry 4.0 relationships and implications. *Sustainability (Switzerland), 13*(6). https://doi.org/10.3390/su13063107

Fraga-Lamas, P., Lopes, S. I., & Fernández-Caramés, T. M. (2021). Green IoT and Edge AI as Key Technological Enablers for a Sustainable Digital Transition towards a Smart Circular Economy: An Industry 5.0 Use Case. *Sensors, 21*(17), 5745. https://doi.org/10.3390/s21175745

Fuentes, S., Tongson, E., & Viejo, C. G. (2024). Artificial intelligence and Big Data revolution in the agrifood sector. In *Elsevier eBooks* (pp. 171–188). https://doi.org/10.1016/b978-0-443-15516-1.00009-8

Gutierrez, A., Boukrami, E., & Lumsden, R. (2015). Technological, organisational and environmental factors influencing managers' decision to adopt cloud computing in the UK. *Journal of Enterprise Information Management, 28*(6), 788–807. https://doi.org/10.1108/jeim-01-2015-0001

Hangl, J., Behrens, V. J., & Krause, S. (2022). Barriers, drivers, and social considerations for AI adoption in supply chain management: A tertiary study. *Logistics, 6*(3), 63.

Hoosain, M. S., Paul, B. S., & Ramakrishna, S. (2020). The impact of 4ir digital technologies and circular thinking on the United Nations sustainable development goals. *Sustainability (Switzerland), 12*(23), 1–16. https://doi.org/10.3390/su122310143

Hsu, C., Wu, J., Zhang, T., & Chen, J. (2024). Deploying Industry 5.0 drivers to enhance sustainable supply chain risk resilience. *International Journal of Sustainable Engineering, 17*(1), 1–28. https://doi.org/10.1080/19397038.2024.2327381

Hu, S. J., & Chen, A. (2023). Unlocking the potential of sustainability: The influence of green innovation and supply chain management on corporate performance. *Environmental Science and Pollution Research, 30*(44), 99774–99788. https://doi.org/10.1007/s11356-023-29022-6

Khan, M., Haleem, A., & Javaid, M. (2023). Changes and improvements in Industry 5.0: A strategic approach to overcome the challenges of Industry 4.0. Green Technologies and *Sustainability, 1*(2), 100020. https://doi.org/10.1016/j.grets.2023.100020

Krstić, M., Agnusdei, G. P., Tadić, S., & Miglietta, P. P. (2023). Prioritization of e-traceability drivers in the agri-food supply chains. *Agricultural and Food Economics, 11*(1). https://doi.org/10.1186/s40100-023-00284-5

Lahane, S., Paliwal, V., & Kant, R. (2023). Evaluation and ranking of solutions to overcome the barriers of Industry 4.0 enabled sustainable food supply chain adoption. *Cleaner Logistics and Supply Chain, 8*, 100116. https://doi.org/10.1016/j.clscn.2023.100116

Lisitsyn, A. B., Chernukha, I. M., & Nikitina, M. A. (2024). Cyber-physical systems in food production chain. *Teoriâ I Praktika Pererabotki Mâsa, 8*(4), 316–325. https://doi.org/10.21323/2414-438x-2023-8-4-316-325

Lv, Z. (2023). Digital Twins in Industry 5.0. *Research,* 6. https://doi.org/10.34133/research.0071

Madhavan, M., Wangtueai, S., Sharafuddin, M. A., & Chaichana, T. (2022). The Precipitative Effects of Pandemic on Open Innovation of SMEs: A Scientometrics and Systematic Review of Industry 4.0 and Industry 5.0. *Journal of Open Innovation: Technology, Market, and Complexity, 8*(3), 152. https://doi.org/10.3390/joitmc8030152

Mandal, S., Kar, A. K., Gupta, S., & Sivarajah, U. (2023). Achieving food supply chain resilience during natural disasters through Industry 5.0 Enablers—Empirical Insights based on an FSQCA approach. *Information Systems Frontiers.* https://doi.org/10.1007/s10796-023-10439-w

Massaro, A. (2023). Advanced electronic and optoelectronic sensors, applications, modelling and industry 5.0 perspectives. *Applied Sciences. 13.* pp.4582. https://doi.org/10.3390/app13074582

Mekkunnel, F. (2019). Industry 5.0: man-machine revolution [Master Thesis, Technische Universität Wien]. repositUm. https://doi.org/10.34726/hss.2019.65408

Mohammed, A. S. (2023). Using Statistical Approach to Derive Priority Weights in the Analytic Hierarchy Process with Application to Public Investment Allocation. *Al-MağAllah Al-MiṣRiyyaẗ Lil Tanmiyyaẗ Wa Al-TaẖṬĪṬ/Al-MağAllah Al-MiṣRiyyaẗ Lil Tanmiyyaẗ Wa Al-TaẖṬĪṬ, 31*(1), 31–53. https://doi.org/10.21608/inp.2023.300183

Mondal, S. K., Banik, S. C., & Rabbi, M. S. (2023). A Fuzzy-AHP analysis to the determination of weights of the main obstacles of RMG in Industry 4.0 application for Bangladesh. *American Journal of IR 4 0 and Beyond*, *2*(2), 1–12. https://doi.org/10.54536/ajirb.v2i2.1728

Nahavandi, S. (2019). Industry 5.0—A human-centric solution. Sustainability, *11*(16), 4371.

Özdemir, V., & Hekim, N. (2018). Birth of Industry 5.0: Making Sense of Big Data with Artificial Intelligence, "The Internet of Things" and Next-Generation Technology Policy. Omics a *Journal of Integrative Biology, 22*(1), 65–76. https://doi.org/10.1089/omi.2017.0194

Poongothai, P., Nandakumar C. D., & Ponnusamy R (2023) Decision-Making for Electric Vehicle Selection Using AHP-MABAC-Fuzzy Integration. In: *2023 International Conference on Data Science, Agents & Artificial Intelligence (ICDSAAI)*, Chennai, India, 2023, pp. 1–5. https://doi.org/10.1109/ICDSAAI59313.2023.10452507

Paschek, D., Mocan, A., & Draghici, A. (2019). Industry 5. 0—The expected impact of next industrial revolution. *Proceeding of Management, Knowledge and Learning International Conference, 2019*, 125–132.

Pomerol, J.-C., & Barba-Romero, S. (2000) Multicriterion decision in practice. In *International series in management science/operations research/International series in operations research & management science*, pp. 299–326. https://doi.org/10.1007/978-1-4615-4459-3_11

Rahimifard, S., Brewer, S., Garcia-Garcia, G., & Jagtap, S. (2022). Digitalising food manufacturing. *Food Science & Technology, 36*(3), 24–27. https://doi.org/10.1002/fsat.3603_5.x

Rahmani, R., Jesus, C., & Lopes, S. I. (2024). Implementations of digital transformation and digital twins: Exploring the factory of the future. *Processes, 12*(4), 787. https://doi.org/10.3390/pr12040787

Rüßmann, M., Lorenz, M., Gerbert, P., Waldner, M., Justus, J., Engel, P., & Harnisch, M. (2015). Industry 4.0: The future of productivity and growth in manufacturing industries; Boston consulting group: Boston. *MA, USA, 2015*, 1–14.

Saifullah, S. (2019). Fuzzy-AHP approach using Normalized Decision Matrix on Tourism Trend Ranking based-on Social Media. *Jurnal Informatika, 13*(2), 16. https://doi.org/10.26555/jifo.v13i2.a15268

Singh, R., Khan, S., Dsilva, J., & Centobelli, P. (2023). Blockchain integrated IoT for food supply chain: A grey based delphi-DEMATEL approach. *Applied Sciences, 13*(2), 1079. https://doi.org/10.3390/app13021079

Skobelev, Petr & Borovik, Sergey. (2021). On the way from Industry 4.0 to Industry 5.0: from digital manufacturing to digital society. 2. 307–311.

Song, J. M., Sung, J., & Park, T. (2019). Applications of blockchain to improve supply chain traceability. *Procedia Computer Science, 162*, 119–122.

Sopha, B. M., Asih, A. M. S., Pradana, F. D., Gunawan, H. E., & Karuniawati, Y. (2016). Urban distribution center location. *International Journal of Engineering Business Management, 8*, 184797901667837. https://doi.org/10.1177/1847979016678371

Varriale, V., Cammarano, A., Michelino, F., & Caputo, M. (2023). Industry 5.0 and triple bottom line approach in supply chain management: the State-of-the-Art. *Sustainability, 15*(7), 5712. https://doi.org/10.3390/su15075712

Vasanthraj, N., Kaur, A., Potdar, V., & Agrawal, H. (2023). Industry 4.0 adoption in food supply chain to improve visibility and operational efficiency—A content analysis. *IEEE Access, 11*, 73922–73958. https://doi.org/10.1109/access.2023.3295780

Wang, M., & Pan, X. (2022). Drivers of artificial intelligence and their effects on supply chain resilience and performance: An empirical analysis on an emerging market. *Sustainability, 14*(24), 16836.

Wang, Z., Chen, Z., Xiao, L., Su, Q., Govindan, K., & Skibniewski, M. J. (2023). Blockchain adoption in sustainable supply chains for Industry 5.0: A multistakeholder perspective. *Journal of Innovation & Knowledge/Journal of Innovation and Knowledge, 8*(4), 100425. https://doi.org/10.1016/j.jik.2023.100425

Wiwatkajornsak, S., & Phuaksaman, C. (2024). Fuzzy AHP-Based evaluation of key success factors in digital marketing for the food retail industry. In *Lecture notes in networks and systems* (pp. 73–84). https://doi.org/10.1007/978-3-031-58561-6_8

Wongvilaisakul, W., Netinant, P., & Rukhiran, M. (2023). Dynamic multi-criteria decision making of graduate admission recommender system: AHP and Fuzzy AHP approaches. *Sustainability, 15*(12), 9758. https://doi.org/10.3390/su15129758

Xiao, G., Samian, N., Faizal, M., Asad, M., Fadzil, M. H. A., Abdullah, A., Seah, W. K., Ishak, M., & Hermadi, I. (2023). A framework for blockchain and internet of things integration in improving food security in the food supply chain. *Journal of Advanced Research in Applied Sciences and Engineering Technology, 34*(1), 24–37. https://doi.org/10.37934/araset.34.1.2437

Zhu, C., Leung, V. C. M., Shu, L., & Ngai, E. C. (2015). Green internet of things for smart world. *IEEE Access, 3*, 2151–2162. https://doi.org/10.1109/access.2015.2497312

Zhang, L., & Su, W. (2023). Corporate social responsibility, internal control, and firm financial performance. *Frontiers in Psychology, 13*. https://doi.org/10.3389/fpsyg.2022.977996

Capital Market in India and Regulatory Tussle Between SEBI and SAT

Milan Prajapati and Manisha Patawari

Abstract This user manual provides instructions on how to use the Springer Nature Word template effectively. It is intended as a guide for Springer Nature book and chapter authors in understanding how the template functions. This user guide has been prepared in the form of a sample chapter which at the same time explains the different features offered by this tool. The concept of stock trading in India can be dated back to the eighteenth century. It was only in 1925 that first legislation to govern security trading was passed, Bombay Securities Contracts Control Act, 1925 ("BSCC Act") with an objective to safeguard public interest. Though this Act failed miserably in achieving its objective. With independence came the legislation governing Securities in the Indian Capital markets, however there were a lot of aspects that had been untouched by the said legislation. With the implementation of the Liberalization and Globalization policy, India became a hub for foreign direct investments and Indian companies also started raising more capital through the markets in order to join the race of the developed economies. There was no such governing body until 1988 which could regulate the entire security market in India. However, with the formulation of the Securities and Exchange Board of India in 1988 and Securities and Exchange Board of India Act in 1992, a lot of revolutionary change came into the legal framework of different aspects of the financial market. Government very smartly slipped in the provision of formulation of Securities Appellate Tribunal in order to keep a check on the powers of the SEBI. With time, SAT has kept on intervening in instances where SEBI has overstepped while adjudicating a matter. This present paper focuses on the scam which took place in the capital markets in the last decade and decoding instances where both SEBI through formulation of different regulations and SAT by assessing the orders of the SEBI has intervened in

M. Prajapati · M. Patawari (✉)
Institute of Law, Nirma University, Ahmedabad, Gujarat, India
e-mail: manisha.patawari@nirmauni.ac.in

M. Prajapati
e-mail: milaneklavya@gmail.com

M. Prajapati
London School of Economics, London, United Kingdom

R. S. Mor et al. (eds.), *Sustainable Business Management, Innovation and Technology*, Sustainable Economy and Ecotechnology, https://doi.org/10.1007/978-3-031-98141-8_9

the domain of the capital market to ensure the trust of the investors and protect their funds.

Keywords Indian Stock exchange · Brokers · SEBI · SAT · Scam

1 Introduction

Capital is a very crucial component of determining the economic growth of a nation. Any financial system that needs to function smoothly has to be first properly structured in order to facilitate sufficient capital generation through different sources like savings, financing, and investments. The capital markets can be understood as a chain of systems in order to generate capital from the investors in the economy through a chain of interconnected subsystems, each consisting of several interdependent parts. These elements work together to achieve a certain outcome within a defined limit. India has a long history of trading. In India, the trading in securities started in the early eighteenth century, but the market was organised for the first time by the Bombay Securities Contracts Control Act, 1925 ("BSCC Act"). This Act was passed with the intention of safeguarding public interest as it is the hard earned savings of the public which is involved in the capital market. Though the new law has failed in achieving its objectives. After independence, India realised the importance of a special law to govern securities contracts, and therefore, in 1956, the Indian parliament enacted the Securities Contract Regulation Act, 1956 (hereinafter referred to as SCRA, 1995) with the objective of solving three major issues of the then capital market: "to regulate the transaction in the securities market, to check nasty speculations, and to regulate buying and selling of securities out of the stock exchange." The Indian securities market was very speculative in the next four decades. As a result of which, the central government established a non-statutory body named the Securities Exchange Board of India in 1988 to advise the Government of India with regard to investor protection and market regulations. In 1991, the Indian market was open to the world; so is our securities market. A lot of companies got foreign funding, and thus the share market was also affected positively by this. This has given opportunities to many stock brokers and investors and therefore also became a time where it was very easy to create a bubble. In 1992, India's largest stock market manipulation was brought into the spotlight, "the Harshad Mehta Scam." In this scam, thousands of investors have lost their lifelong savings. This incident became a turning point in the history of the Indian Securities Market and gave new responsibilities on the shoulders of SEBI. In 1992, the Indian parliament passed the Securities Exchange Board of India Act.

2 Literature Review

In a paper entitled "Lux industry limited: A case study (Sanjay Dhamiji, 2023), discusses the case of lux limited. The article describes the role of SEBI in case of financial fraud. In the present case the promoter and the promoter group of Lux Industry Ltd. were alleged of insider trading through a chain of 14 companies. This paper highlighted the role of the regulator in tackling the issues of insider trading and ensuring that investor confidence remains intact in the market.

The study focused on revealing various illegal and dubious practices that brokers in the stock exchange adopt to deceive investors. The study further discusses the use of technology for robust plugging of these loopholes in the current market.

In a paper titled "The SEBI Role in Indian Stock Market" (Dasari Rajesh Babu, 2019), the author talks about how the stock markets in India are overseen by the Securities and Exchange Board of India (SEBI), which was founded under the SEBI Act of 1992. The author also discusses that SEBI's key responsibilities include regulating and developing the Indian securities markets and looking out for investors' best interests. The paper further focuses on investor protection regulations, present market tendencies, and the evolution of these policies over the past twenty years.

In another paper titled "Regulation and Deregulation of the Stock Market in India (Goyal, 2005), the author talks about how globalisation renders government limitations ineffective and new technologies need new governance structures; therefore, liberalization entails de- and re-regulation. The redesigned Indian regulatory system is evaluated using fundamental regulatory ideas, capital market regulatory requirements, and Indian market features. The regulator aimed at developing global best practices, market integrity, and self-enforcing game regulations while also supporting the game. This facilitates the market adoption of world-class technologies and practices.

One of the papers titled "Securities and Exchange Board of India and the Indian Capital Markets: A Survey of the Regulatory Provisions (Sabarinathan, 2004) talks about how SEBI has become a prominent presence in the Indian capital market due to the significant expansion of the market and its various efforts to regulate its operations, safeguard public investors, and foster the creation of new financial products and institutions. Although investors, scholars, and ordinary people have a basic understanding of the function that SEBI is required to fulfil and has fulfilled, there is limited comprehension of the specific legislative mechanisms via which SEBI operates, particularly from an economic one.

In a column entitled "Imposition of costs: A tussle between SAT and SEBI," the author has argued that time and again SAT has overruled the order of SEBI and, in many cases, has passed adverse remarks on the order passed by SEBI. In the present article, the author has discussed various cases where SAT has explicitly talked about gross negligence by the investigating officers of SEBI (Kataria, 2022).

3 Research Objective

To study the evolution of the securities market in the last decade.

- To analyse the role of SEBI as regulator of the capital market.
- To scrutinise how the investors are defrauded in the capital market even after the establishment of SEBI.
- To investigate the role of SAT in balancing the power dynamics between SEBI and other stakeholders in the market.

4 Research Questions

1. How have the capital markets in India been evolving with the implementation of different legislation and regulations with the changing times?
2. What role does SEBI have as a regulator in the capital markets in India?
3. Whether intermediaries and companies still manage to defraud investors in the capital markets in India?
4. How has the Securities Appellate Tribunal kept a check on the interventions on SEBI in different cases in recent times?

5 Methodology

In the present paper the authors have followed doctrinal methods of research. The judicial decisions were analysed to understand the power dynamics between the two quasi-judicial institutions, "SEBI and SAT.

5.1 Role of SEBI in Capital Market

The Securities Exchange and Board of India ("SEBI") was established by the Indian government in 1988 as a non-statutory body (Stock and Exchange Board of India, n.d.) against the already existing Act on Capital Issues (Latest Law, n.d.). Before the Capital Issue (Control) Act, the securities were dealt through the Defence of India Rules in 1943 under the Defence of India Act (Defence of India Act, 1939). As per the provisions of the Capital Issue 3 Act, companies were required to obtain approval from the central government before issuing capital (Capital Control Act, s.4). SEBI was formed with an intent to facilitate the trading of corporate securities on various stock exchanges throughout India. Since the investors during the times of financial instability and political turmoil going on in the country had lost faith in the financial markets in India, SEBI played a role in promoting investor confidence and protecting the rights of investors. In 1992, when the liberalisation policy was supposed to be implemented, the P.V. Narasimha Rao government decided to give SEBI statutory power by implementation of the SEBI Act of 1992. The stock market saw several transformations throughout the 1980s. There has been continuous growth in investor engagement and market expansion. There are now around 28 stock exchanges in India that are listed under SEBI's supervision (SEBI Details of stock exchange, n.d.). The stock market serves as a venue where investors and traders may engage in the buying and selling of shares and mutual funds. Within the financial system, there are several financial market intermediaries that play a major role in easing out the work of the investors willing to invest in the financial markets. These individuals are alternatively referred to as merchant bankers or investment bankers and regularised through different regulations of SEBI (SEBI (Merchant Banker Regulation) 1992, n.d.) [Last Amended on August 18, 2023]. The sole purpose of these intermediaries is to facilitate the investment process and build a connection between investors and fund users. According to the SEBI Act, individuals and organisations involved in the securities market, such as stockbrokers and sub-brokers (SEBI (Stock Broker) Regulation, 1992, 2023), are only allowed to purchase, sell, or trade securities if they meet the requirements stated in their registration certificate. SEBI periodically arranges and implements investor awareness initiatives through the organisation of many seminars, workshops, and conferences. SEBI has established a specially dedicated website to provide investors with comprehensive information and address their issues. Looking at the growth of capital markets in India, the government, under the powers given to it under the SEBI Act, has made regulatory interventions in the capital markets through the drafting of several legislations and regulations over a period of time in order to regulate the capital market transactions:

1. In 2015, another regulation was released to prohibit the insider trading practices generally practiced by the stock and sub-stock brokers in different types of financial markets in India. Apart from prohibiting the insider trade practices, this regulation aimed at creating a framework and legally strengthening it.
2. Intermediaries play a very major role in the capital market, and over the period of time a lot of these intermediaries have been wrongly using the investor money due to any regulation. Finally, the SEBI, through an official notification, released regulations to govern these intermediaries in the market in 2008. (SEBI (Intermediaries) Regulation, 2008), 2022)
3. The goal of KYC requirements is to prevent criminals from using banks for money laundering purposes. It also allows banks to better understand their clients and their financial transactions in order to service them better and manage risks responsibly. Looking at the various frauds in the capital markets, SEBI introduced a regulation for getting proper information about the client through the KYCS Securities Registration Agency ("KRA") (SEBI (KYC Registration Agency) Regulation, 2011, 2024).
4. Stock brokers are regulated professionals, typically affiliated with a brokerage firm or broker-dealer, who engage in the buying and selling of stocks and other securities for both retail and institutional clients, either through a stock exchange

or over the counter, in exchange for a fee or commission. Since all the securities in the capital market are traded by them, SEBI made a regulation SEBI (Stock brokers & sub-brokers) Regulations, 1992 to govern the stock brokers and sub-stock brokers who act as their agents. All these brokers before trading are mandated to register with the SEBI first.

6 Scams in Capital Market in Last Two Decades

It has been observed now that for a very long time, the public's preferred investing option has been the securities market, where investors prefer to trade in different kinds of securities. This investment practice has brought in a lot of wealth for the investors, resulting in high returns. However, this fact cannot be ignored: though there have been so many new regulations brought in to keep a check on companies from committing any fraud, there have been instances in which scammers have deceived investors and the market, resulting in irrevocable and irreplaceable harm.

The Shruti Vohra WhatsApp leak case (Shruti Vohra vs. Securities and Exchange Board of India, 2021) was one of the major corporate fraud cases as a result of loopholes in insider trading regulations. India already had the SEBI Act1 and the PIT Regulations of 1992, which addressed the issue of insider trading. However, with the change in time, SEBI understood the gravity of this issue and released the new PIT Regulations of 2015. The new regulations have dealt with the issue of communicating the unpublished price sensitive information ("UPSI") under Sect. 3, which talk about the communication or procurement of unpublished price sensitive information, and Sect. 4, which talks about trading when in possession of unpublished price sensitive information of the PIT regulations. The maximum penalty for committing the crime of contravening with these provisions, SEBI regarding USPI, should not be less than 10 lakh rupees, and the maximum is Rs 25 crores, or three times the gains from insider trading, whichever is higher, as per the Companies Act of 2013 and SEBI Act. In this case, price-sensitive information regarding huge stock market giants like Wipro, Mindtree, Bajaj Auto, etc. was being circulated through different WhatsApp channels. The accused had been contending that this circulation of messages is a common practice amongst the traders where a concept of Heard on Street ("HOS) is followed, which means false whispers or gossip that are being frequently circulated as speculation by brokerage firm analysts. However, later on, the Securities Appellate Tribunal ("SAT") heard an appeal against the ruling of SEBI made by the employees of the broking firm, which were accused in this case. Following careful examination, SAT overturned SEBI's ruling that the employees were insiders for sharing price-sensitive information. In support of its ruling, SAT stated that SEBI was unable to identify the source of this financial disaster. Thus, it is inappropriate to hold anyone accountable for nothing more than simply transmitting a communication.

In the case of "Manoj Gokulchand Seksaria vs. The State of Maharashtra" on 5 January 2022 at Bombay High Court, several brokers and employees of Yes Bank Limited were named in the First Information Reports ("FIRs") because of allegations that they participated in the illegal allocation of IPOs under false identities. The goal was to unfairly benefit from the initial public offerings ("IPOs") by purchasing more shares than were intended for individual investors and then selling them at a higher price. Corporations would be opening Demat accounts under fictitious or benami names so they could control the distribution of most of the shares in IPOs. On the other hand, a circular describing guidelines for consent orders and the assessment of petitions to resolve the breaches was issued by SEBI in April 2007. The SEBI is empowered to resolve both civil and criminal cases that it has brought or is investigating against individuals or entities that are believed to have violated SEBI statutes, according to these regulations. During the proceedings, the court concluded that none of the legitimate investors incurred any financial harm due to the alleged activities, which had some commercial implications. Section 24-A of the SEBI Act, which talks about the composition of certain offences, grants the SEBI the authority to prosecute offences at any stage of the process. The court concluded that the petitioner's ongoing prosecution was deemed an abuse of process due to his strategic utilisation of the composite plan, which ultimately undermined the integrity of the whole case. Basically, the Court's analysis showed that the SEBI could bring the same charges against anyone who has broken the rules in both criminal and civil proceedings. In this case, prior to finalising the charges, the SEBI considered both the substance of the offence and the broader public interest. (Manoj Gokulchand Seksaria versus The State of Maharashtra, 2022).

In 2009, the National Stock Exchange ("NSE") introduced the system of co-location services, which allowed the brokers and traders registered with the stock exchange to lease space in their data centre for the operation of their servers. Due to their close proximity with the stock exchange computers, they were able to quickly access the price information for the traders who logged in first. In 2015, one of the whistleblowers from the stock exchange made many accusations to the market regulator regarding this malpractice followed by the brokers. These charges asserted that certain algorithmic traders possessed technical specifications that enabled them to access NSE systems and the exchange's data stream at a slightly faster pace than their competitors, with a difference of only a fraction of a second. The system of "Tick-By-Tick" ("TBT") data feed releases information in a sequential manner based on the order in which brokers joined or signed in to the server. When the matter reached SEBI, its inquiry revealed that the NSE's methods and procedures were unjust, lacked transparency, and showed bias, thereby violating the terms of the regulations on Stock Brokers and the SEBI ("PFUTP"). The SEBI, through an order dated April 30, 2019 (ORDER No. WTM/GM/EFD/03/2018–19 under Sects. 11, 11(4), and 11B of the SEBI Act, 1992), has levied a penalty of Rs. 625 crore on the National Stock Exchange (NSE) for its inability to guarantee equitable access to its trading systems. In addition, the regulatory authority has prohibited the exchange from introducing any new goods or services for a period of six months. Furthermore, it has instructed the exchange to carry out a comprehensive examination of its systems and procedures. The NSE was instructed to establish effective methods and procedures to provide equitable access to its trading systems.

7.4. In the Karvy Stock Broking Limited ("KSBL") scam, KSBL was a stock-broker who controlled the securities of its clients. KSBL made a smart move by misusing the Power of Attorney ("PoA") given by its clients to secure loans. They used these securities as collateral and showed it as their own known securities while securing the loan amount. This misuse of the securities took a severe toll on the clients trust, since investors were never informed about this and they simply were under the notion that the broker would be acting responsibly and transparently. By pledging these securities, KSBL violated several provisions of the SEBI regulations 2. This unauthorized pledging of these securities posed a serious threat to the investors assets. When the matter reached SEBI, it issued a circular (SEBI Circular No.: CIR/HO/MIRSD/DOP/CIR/P/2019/75) to bar the practice of the stock brokers of pledging the securities of clients in the procurement of loans. Once the investigation was initiated by SEBI, the National Stock Exchange ("NSE") reported that approximately out of the 95,000 client's of KSBL, 83,000 clients securities had been pledged, and the valuation of the scam summed up to Rs. 2300 crores. Looking at this reported data, SEBI passed an order to stop KSBL from taking any more clients and also ordered National Stock Depository Limited ("NSDL") to allow KSBL to transfer the securities to those clients who had fully paid up for those securities. After looking at this scam, SEBI wanted to strengthen the regulation to put an end to this loophole in the Intermediaries Regulation. SEBI recently released a circular dated April 25, 2023, (SEBI Circular No.: (SEBI/HO/MIRSD/MIRSD-PoD-1/P/CIR /2023/061) in order to put a restriction on the stock brokers from pledging the securities and money of the clients.

7 Intervention Made by Securities Appellant Tribunal

As per the provisions of the SEBI Act, the Securities Appellate Tribunals ("SAT") have been formulated for challenging the orders or fines issued by SEBI, ensuring access to appropriate appellate remedies. The SAT acts have a quasi-judicial body, and it is the most unique quasi-judicial body since it has a special power to review its own order (SEBI, 1992, Sec. 15U). As per the SEBI Act, the Securities Appellate Tribunal will be composed (SEBI 1992, Sec. 15L) of a presiding officer and a specific number of judicial members and technical members, as determined by the Central Government through a notification. These members will have the authority to exercise the powers and fulfil the responsibilities given to the Securities Appellate Tribunal under this Act or any other applicable law. Some of the interventions made by SEBI in recent times are as follows:

After the case of Shruti Vohra versus Securities and Exchange Board of India, SEBI is now taking a serious toll on the information that is being circulated on social media. Any kind of information that falls within the preview of UPSI is being penalised as a violation of the insider trading regulation. SEBI also looking at the previous PIT regulations brought an amendment in 2018 (SEBI PIT (Amendment) Regulations, 2018). To put an end to the brokers taking the defence of HOD, the SEBI, through an order, said that in this case the information that was being circulated was private and was not in the public notion, and the way the information was obtained was not out of the result of any study conducted or a survey (Economic times, 2021).This order of SEBI acts as a tool to measure the HOD practice by the stock brokers.

In the case of the (National Stock Exchange and Ors. versus Securities and Exchange Board of India, 2023) SEBI instructed the exchange to carry out a comprehensive examination of its systems and procedures. The NSE was instructed to establish effective methods and procedures to provide equitable access to its trading systems. The Securities Appellate Tribunal ("SAT") did not find the order of SEBI appropriate since there were a lot of discrepancies in the said case. With regards to the SAT Regulations (Securities Appellate Tribunal (Procedure)), the NSE incurred a penalty of Rs 100 crores for its failure to meet the anticipated requirements as a major regulator. The SAT itself acknowledged the absence of specific criteria to quantify the contraventions of the provisions committed by the NSE but recognised that the magnitude of these infringements is substantial enough to act as a deterrent. SAT finally set aside the order in 2014 and has switched to a multicast TBT system in order to adhere to the order execution in the coalition facility.

The SEBI issued several Show Cause Notices ("SCNs') to PwC in relation to PwC's audit of Satyam Computer Services Limited (Price Waterhouse & Co. v. SEBI, 2010). When the SEBI took this matter into investigation, it discovered that Satyam had misrepresented its current account bank balances, fixed deposit balances, interest income revenue from sales, and debtors' numbers in its books of account and financial statements for several years. SEBI observed that PwC, as the statutory auditors, failed to adequately discharge their responsibilities, resulting in the public being encouraged to continuously trade in the company's shares. The auditors were observed to have made significant claims in their certifications without providing any supporting documentation, indicating a severe lack of care and deliberate fabrication. The SEBI, through an order (SEBI Order No. WTM/GM/DRA 1/83 /2017–18), put restrictions on the auditing services of PwC, and in return, PwC lodged an appeal against the SEBI ruling before the SAT. Nevertheless, the SAT, in a ruling, declined to issue a stay on the SEBI order. However, after listening to the arguments, the SAT bench noted that the SEBI order will not hinder PwC and its network firms from carrying out audit assignments for the fiscal year 2017–18. It further clarified that the SEBI order will not affect the audit assignments that PwC has already undertaken for clients who follow the financial year starting on January 1, 2018. (Upadhyay, 2018).

KSBL, dissatisfied with the ex-parte ad-interim order issued by SEBI, has lodged an appeal with SAT seeking clarity. As a result of this ex-parte ad-interim ruling, KSBL is experiencing challenges in resolving the transactions of its clients with the clearing house. The lenders of KSBL, including HDFC Bank, ICICI Bank, Bajaj Finance Limited, and IndusInd Bank, lodged separate appeals with the Securities Appellate Tribunal (SAT) in response to the SEBI ruling. They also desired the SAT to overturn the NSDL's decision to transfer these stocks to customer accounts. They claimed that they possessed legitimate rights over such securities and that it was a standard practice in the business to lend money against pledged shares. Axis Bank argued that the pledge made by KSBL on the securities complied with the regulations stated in the SEBI circular (SEBI Circulars No. SEBI/HO/MIRSD/MIRSD2/CIR/P/2016/95) and was a legally binding pledge. Furthermore, it has been said that a pledge that has been properly made is not invalidated by the SEBI circular of 2019 (SEBI Circular No. CIR/HO/MIRSD/DOP/CIR/P/2019/75). This circular only imposes a duty on stock brokers to unpledged all stocks belonging to their clients and does not deem the previously created pledges as invalid. After hearing the appeals, the SAT had granted permission to Axis Bank to exercise its rights over the pledged shares and has instructed the SEBI to reinstate the promise in favour of the other banks within four weeks, along with providing them with compensation, since it was a commercial bank and not a registered trading member with the SEBI. (Aggarwal, 2024) (Times et al., Sat quashes NSE's directive to Axis Bank in karvy case).

8 Conclusion

The ongoing disputes between the Securities and Exchange Board of India (SEBI) and the Securities Appellate Tribunal (SAT) highlight a rising tension between regulatory enforcement and judicial oversight. SAT's criticism of SEBI's "mechanical orders" and excessive penalties raises concerns about procedural fairness, regulatory overreach, and the necessity for a more balanced approach in enforcing securities laws. While SEBI is vital for maintaining market integrity, its often rigid and punitive methods have attracted judicial scrutiny, prompting questions about accountability and transparency. The call for a Uniform Securities Code has become clear, as it could simplify regulatory processes, impose checks on SEBI's power, and ensure fairer adjudication. A well-designed legal framework would not only boost investor confidence but also create a more predictable and equitable regulatory environment. Looking ahead, cooperation among SEBI, SAT, and the government will be crucial in establishing a regulatory ecosystem that is both strong and fair. Rather than seeing SAT's interventions as obstacles, SEBI should view them as a chance to improve its enforcement strategies and develop a framework that emphasizes transparency, proportionality, and due process.

References

Aggarwal, S. (2024, December 13). SAT imposes costs on SEBI for passing another mechanical order: Need for a uniform securities code. Retrieved from Bar and Bench: https://www.barandbench.com/columns/costs-imposed-on-sebi-by-sat-for-passing-another-mechanical

Babu, D. R. (2019). The SEBI role in Indian stock market. 8th International conference on digital outreach and future of management practices—2019 International Journal for Research in Engineering Application & Management, 90.

Economic times. (2021, 11 30). Retrieved March 28, 2025. From https://economictimes.indiatimes.com/markets/stocks/news/sat-quashes-nses-directive-to-axis-bank-in-karvy-case/articleshow/88011928.cms

Goyal, A. (2005). Regulation and De-Regulation of the Stock Market in India. SSRN, 30.
Kataria, S. A. (2022, May 12). Monaq.com. Retrieved March 28, 2025. from https://www.mondaq.com/india/antitrust-eu-competition/1192468/imposition-of-costs-on-sebi-a-tussle-between-sat-and-sebi
Lakshmi, K. A. (2024). An analysis of stockbroking frauds and regulatory action in India. Journal of financial Crime, 1037.
Latest Law. (n.d.). Retrieved March 28, 2025, from https://www.latestlaws.com/bare-acts/central-acts-rules/banking-laws/capital-issues-control-act-1947
Manoj Gokulchand Seksaria vs. The State of Maharashtra, WRIT PETITION NO. 245 OF 2020 (Bombay High Court January 5, 2022).
National Stock Exchange and Ors. vs. Securities and Exchange Board of India, Appeal No. 445 of 2022 (Securities Appellate Tribunal December 14, 2023).
Price Waterhouse & Co. SEBI, WRIT PETITION NO. 5249 OF 2010 (Bombay High Court August 13, 2010).
Sabarinathan, G. (2004, July). Repository- iimb. Retrieved March 28, 2025, from https://repository.iimb.ac.in/bitstream/123456789/468/1/WP.IIMB.228.pdf
Saif, S. (2008). A Study of Stock Market Investors of Delhi—A Behavioural Finance Perspective. Indian Journal of Finance, 24.
Sanjay Dhamija, R. N. (2023). Lux Industries Limited: a case of insider trading. Emerald Emerging Markets Case Studies, 1.
SEBI (Intermediaries) Regulation, 2008). (2022, August 1). Retrieved March 28, 2025, from https://www.sebi.gov.in/legal/regulations/aug-2022/securities-and-exchange-board-of-india-intermediaries-regulations-2008-last-amended-on-august-1-2022-_61700.html
SEBI (KYC Registration Agency) Regulation, 2011. (2024, November 28). Retrieved March 28, 2025, from https://www.sebi.gov.in/legal/regulations/nov-2024/securities-and-exchange-board-of-india-kyc-know-your-client-registration-agency-regulations-2011-last-amended-on-november-28-2024-_89274.html
SEBI (Merchant Banker Regulation) 1992. (n.d.). Retrieved March 28, 2025, from https://www.sebi.gov.in/legal/regulations/aug-2021/securities-and-exchange-board-of-india-merchant-bankers-regulations-1992-last-amended-on-august-03-2021-_35135.html
SEBI (Stock Broker) Regulation, 1992. (2023, August 18). Retrieved March 28, 2025. from https://www.sebi.gov.in/legal/regulations/aug-2023/securities-and-exchange-board-of-india-stock-brokers-regulations-1992-last-amended-on-august-18-2023-_76325.html
SEBI Details of stock exchange. (n.d.). Retrieved March 2025. From https://www.sebi.gov.in/stock-exchanges.html
Shruti Vohra. (2021). Securities and Exchange Board of India, CA No.2252-2262. 202 (Supreme Court of India March 22, 2021).
Stock and Exchange Board of India. (n.d.). Retrieved March 28, 2025. From https://www.sebi.gov.in/about-sebi.html
Upadhyay, J. P. (2018). SAT refuses to lift SEBI ban on Price Waterhouse. Retrieved from Livemint: https://www.livemint.com/Money/WgbGqcNWDuGhwwrMIsR1lN/SAT-refuses-to-lift-Sebi-ban-on-Price-Waterhouse.html

Balancing the Triple Bottom Line: Analysing Sustainable Dynamics Through TISM Analysis of Environmental, Social, Economic, and Ethical Factors in Achieving Social Development Goals

Shubhangi Tyagi, Sushma Sharma, Anju Rohilla, and Deepti

Abstract The aim of this study is to identify key sustainability factors across environmental, social, economic, and ethical domains and to investigate the relationship between these factors. The study applied a systematic literature review and total interpretive structural modeling (TISM) to classify and analyze the factors influencing sustainability dynamic factors. A digraph model of these variables was developed. Also, MICMAC (the fuzzy Matrices' Impacts of Croise's Multiplication Appliquée a UN Classement) analysis was applied to classify the factors into driving and dependent categories. The study identified nine variables that influence sustainability dynamics, emphasizing that Innovation and Technological Advancement and Consumer Awareness are crucial drivers in achieving sustainable dynamics, pointedly manipulating various environmental, social, and economic factors. Moreover, variables for instance, Labour Welfare and Corporate Social Responsibility play key roles in enhancing interdependencies, achieving social development goals. This study offers innovative insights into the interrelation of crucial factors influencing sustainability dynamics, providing a framework aiming to achieve societal development goals. Organizations should prioritize Innovation and Technological Advancement and Consumer Awareness in sustainability strategies while enhancing Corporate Social Responsibility and Labour Welfare for actual progress toward social development goals.

Keywords Sustainability · MICMAC analysis · ISM · Interdependencies · Sustainable development

S. Tyagi · S. Sharma (✉) · A. Rohilla
SRM University, Sonepat, Haryana, India
e-mail: sushma.s@srmuniversity.ac.in

Deepti
Assistant Professor, SRM University, Sonepat, Haryana, India

R. S. Mor et al. (eds.), *Sustainable Business Management, Innovation and Technology*, Sustainable Economy and Ecotechnology,
https://doi.org/10.1007/978-3-031-98141-8_10

1 Introduction

Prior to 1991, the idea of sustainability was limited to the goal of preserving resources like those available through fishing (Lele, 1991). However, it has expanded to include other aspects such as environmental preservation to protect the environment for future generations. In order to achieve the goal, green practices must be used according to Chauhan et al. (2020). The goal of green practices is aimed at achieving 'triple bottom line' i.e. environmental, economic and social benefits (Mangla et al., 2014). As a result, sustainability has to take into consideration the economic, social and environmental domains (Khoshfetrat et al., 2020). Without established approaches, achieving sustainable objectives becomes challenging (Yu et al., 2019). Because of this, in India, the implementation is hampered by issues like costs and awareness of the benefits in the long run (Mangla et al., 2014). In order to promote sustainability, the research employs ISM and MICMAC techniques to analyse the sustainability drivers and their relationships towards promoting effective sustainability measures.

1.1 Sustainable Research

Sustainability in Policy, Business, and Society—In the context of ESE, sustainability is a solution to the global environmental, socio political and economic issues. Torelli (2021) &Bebbington et al. (2007) stated that corporate social responsibility is crucial because they foster social and economic progress, whereas, (Bebbington et al., 2007) argue that sustainability is intertwined with CSR.

SDGs as a Framework for Sustainable Development—The SDGs as a framework introduced in 2015 serve as universally applicable guidelines for sustainable development. Additionally, economic factors were included in the model known as the ECON-ESG by (Işık et al., 2024) whereas (Sheehy & Farneti, 2021) believe that companies may achieve success by complying with the sustainable development goals.

Environmental, Social, Economic, and Ethical Dimensions of Sustainability—According to Govindan et al. (2021), sustainability involves ecological, social, economic and all moral aspects. Moreover, the need for green products and reverse logistics in an environmental perspective was also highlighted. While, Chauhan et al. (2020) explored the equilibrium of profit, society and environment.

System Analysis: ISM and MICMAC—ISM and MICMAC assessment are key methods applied in the study of complex systems. (Bux et al., 2020) and (Nida et al., 2024) mentioned that these are applicable in business where they help to identify the factors that boost sustainability by interlinking various aspects (Fig. 1 and Table 1).

Fig. 1 Key concepts in achieving SDGs: a word cloud representation. *Source*: Compiled by the Authors

Table 1 Key insights in sustainability research

Field of study	Key terms	References
Sustainability in modern policy, Business, and Society	Sustainability, Corporate Social Responsibility (CSR), Environmental, social, Economic	Ayu Bebbington et al. (2007), Purnamawati et al. (2023), Torelli (2021)
SDGs as the framework of sustainable development	Sustainable Development Goals (SDGs), ECON-ESG model, Macroeconomic indicators, GDP	Isik et al. (2024), Sheehy & Farneti (2021)
Environmental, Social, Economic, and Ethical dimensions of sustainability	Green products, Reverse logistics, Social sustainability, Economic sustainability, Ethical sustainability	Chauhan et al. (2020), Govindan et al. (2021)
ISM and MICMAC in system analysis	Interpretative Structural Modeling (ISM), MICMAC analysis, Systems analysis, Sustainability drivers	Bux et al. (2020), Nida et al. (2024)

Source: Compiled by the Authors

1.2 Research Objectives

The objective is to identify key sustainability factors in the environmental, social, economic, and ethical domains, explore their interconnections through ISM and MICMAC, and create a hierarchical model that emphasizes the essential driving and dependent elements for sustainable development.

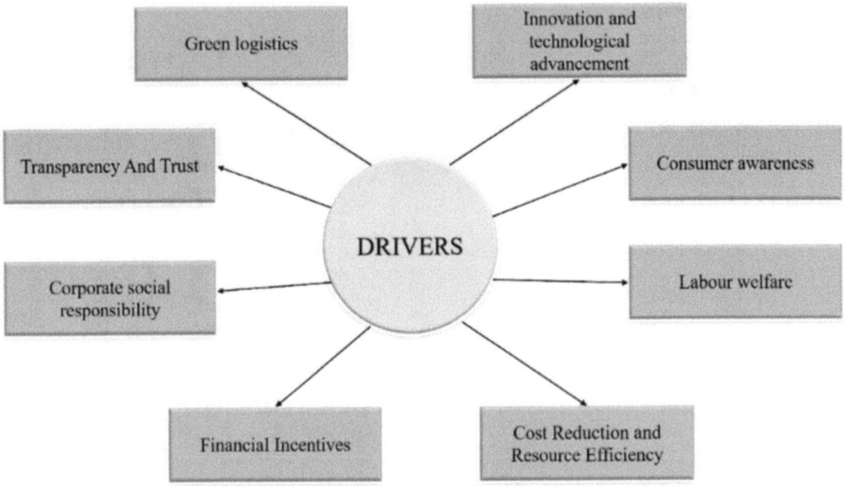

Fig. 2 Drivers of sustainability dynamics. *Source*: Compiled by the Authors

1.3 Research Questions

RQ1: What are the key sustainability factors across environmental, social, economic, and ethical domains, and how do they influence sustainable development?

RQ2: How can ISM and MICMAC methodologies be applied to analyze the interrelationships between these sustainability factors?

RQ3: Which factors act as primary drivers of sustainability, and how do dependent factors contribute to the overall sustainability framework?

Identification of Variable.

1.4 Drivers

See (Fig. 2 and Table 2).

1.5 Variables Used in the Model Construction

See (Table 3).

Table 2 Drivers of sustainability trends

Category	Specific drivers	Description	Sources
Environmental	Green Logistics and Innovation and technological advancement	The aim of green logistics is to minimize the environmental effects of supply chains by employing innovative technologies such as green roofs and energy-efficient buildings to decrease energy consumption and emissions, thus supporting environmental sustainability	Li et al. (2021), Nida et al (2024)
Social	Consumer awareness and Labour welfare	Increasing consumer consciousness about environmental concerns fosters sustainable consumption, while labour welfare practices guarantee equitable pay, human rights, and secure working environments, advancing social sustainability	Ansumensah (2021), Govindan et al. (2021), Hina et al. (2022), Jain & Hudnurkar (2022)
Economic	Cost reduction and resource efficiency and financial incentives	Utilizing methods to reduce costs and improve resource efficiency can help businesses lower operational expenses, while financial benefits like tax breaks and subsidies encourage the use of eco-friendly technologies and sustainable methods	Gardiner & Hajek et al. (2020), Hina et al. (2022), Neves & Marques (2022)
Ethical	Corporate social responsibility and Transparency and Trust	Corporate Social Responsibility incorporates social, environmental, and ethical obligations into business plans to improve sustainability and build trust with stakeholders. Transparency in business practices promotes accountability and builds trust among stakeholders	Ayy Purnamawati et al. (2023), Nida et al. (2024)

Source: Compiled by the Authors

Table 3 The variables used in the study

Variables	Abbreviation
Green logistics	GL
Innovation and technological advancement	ITL
Consumer awareness	CA
Labour welfare	LW
Cost reduction and resource efficiency	CRRE
Financial incentives	FI
Corporate social responsibility	CSR
Transparency and trust	T& T

Source: Compiled by Authors

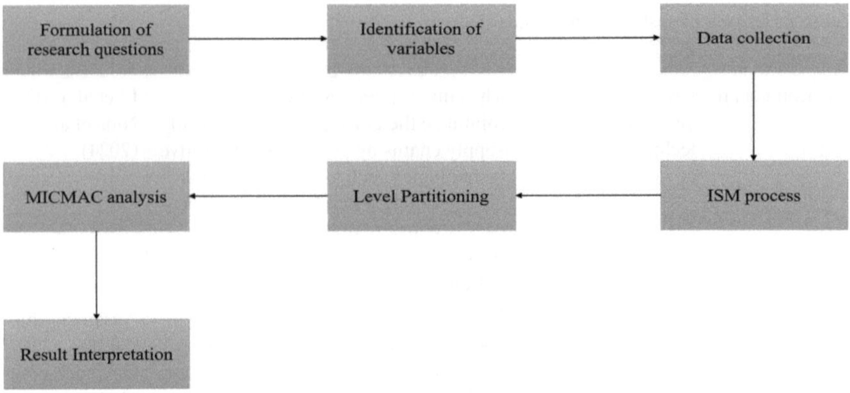

Fig. 3 Process of methodology. *Source:* Compiled by the Authors

2 Methodology

This research adopts the ISM and MICMAC analysis approaches to examine the relationship of the sustainability factors under green logistics, innovation, as well as consumer awareness as key elements. While, ISM presented a model in a hierarchy form, MICMAC pinpointed Innovation and Technological Advancement as the major forces, thus offering direction on how to achieve the Sustainable Development Goals (Fig. 3).

3 ISM Process

The ISM approach studies various components of the system and the interrelationships between them, helping to recognize how complex systems can be broken down into controllable sub-systems and ordered into hierarchies with the aid of decision makers (Vafadarnikjoo et al., 2020). The process comprises determining the critical attributes, developing contextual relationships and creating a SSIM (Structural Self Interaction Matrix), using the symbols V, A, X and O (Bhosale & Kant, 2016). The SSIM is replaced by a reachability matrix whose elements are binary, and transitivity is used. At the end, a directed graph is formed in which an ISM model is created and then verified (Vafadarnikjoo et al., 2020).

Table 4 signify that green logistics directly influences Cost reduction and resource efficiency, while being impacted by Innovation and technological advancement, Consumer Awareness, and Labour Welfare. On the other hand, financial incentives show power mutuality, defining transparency and trust and being exploited by numerous factors. Crucial factors for instance, innovation and technological advancement and corporate social responsibility play a key role in manipulating

Table 4 Structural self-interaction matrix (SSIM)

Elements	GL	ITL	CA	LW	CRRE	FI	CSR	TT
GL		O	A	O	O	A	A	A
ITL			V	V	O	V	O	O
CA				V	V	V	O	V
LW					V	V	V	A
CRRE						A	A	A
FI							V	V
CSR								X
TT								

Source: Compiled by the Authors

other variables in the system, developing the strong foundation for sustainability strategies.

4 Reachability Matrix

Table 5 represents consumer awareness as the most significant driving force, clearly affecting other variables. However, high dependency power of cost reduction and resource efficiency is often reliant on other factors. Other important factors include innovation and technological advancement which are key drivers of critical elements of sustainability.

Table 5 The primary reachability matrix generated by SSIM

Variables	GL	ITL	CA	LW	CRRE	FI	CSR	TT	Driving Power
GL	1	0	0	0	0	0	0	0	1
ITL	0	1	1	1	0	1	0	0	4
CA	1	0	1	1	1	1	0	1	6
LW	0	0	0	1	1	1	1	0	4
CRRE	0	0	0	0	1	0	0	0	1
FI	1	0	0	0	1	1	1	1	5
CSR	1	0	0	0	1	0	1	1	4
TT	1	0	0	1	1	0	1	1	5
Dependency Power	5	1	2	4	6	4	4	4	

Source: Compiled by the Authors

5 Level Partitioning

In ISM analysis, variables for instance, Green Logistics and Cost Reduction and Resource Efficiency are positioned at number 1, representing their maximum reliance on other variables. Power factors such as Innovation and Technological Advancement, Consumer Awareness, Labour Welfare, and others act as crucial factors, pointedly influencing the system. This partitioning simplifies the linkage among these factors, emphasizing which ones to rank for achieving sustainability goals.

6 Final Reachability Matrix

The final reachability matrix is generated, reflecting the comprehensive influence and interconnectedness of the variables within the model. It highlights Innovation and Technological Advancement and Consumer Awareness as the key drivers of sustainability, with the maximum impact on other factors. Meanwhile, Green Logistics and Cost Reduction and Resource Efficiency have complex dependence control, dependent on other variables for influence.

7 Diagraph

See (Fig. 4).

The diagraph from the ISM process stated that variable 1 was the key driver, applying basic influence over the other factors in the system. Factors like 2, 3, 4, 5, 7, and 8, positioned lesser, are more reliant on these complex level drivers. The connections and arrows suggest the direct impact routes, illustrating the interdependencies that outline the system's overall dynamics.

8 MICMAC Analysis

See (Fig. 5).

MICMAC analysis categorizes factors shaped on their driving and dependence powers, with linkage variables for instance, innovation and technological advancement and corporate social responsibility having mutually high impact and dependance, making them complex to the system. reliant on variables like labour welfare dependent on others but have negligible effect, while autonomous factors like green logistics show low powerful and dependance power. This analysis helps prioritize key factors essential for system dynamics.

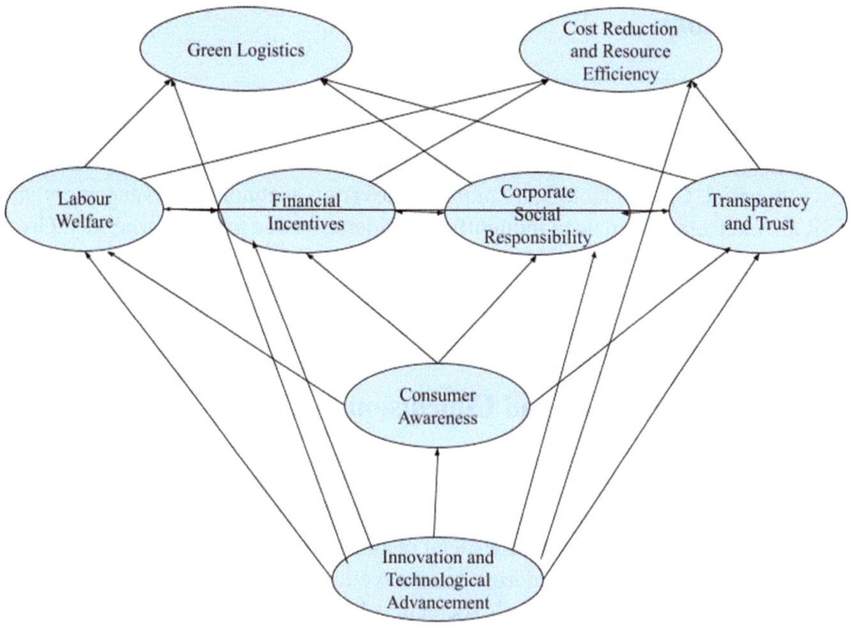

Fig. 4 A diagrammatic representation of sustainability dynamics

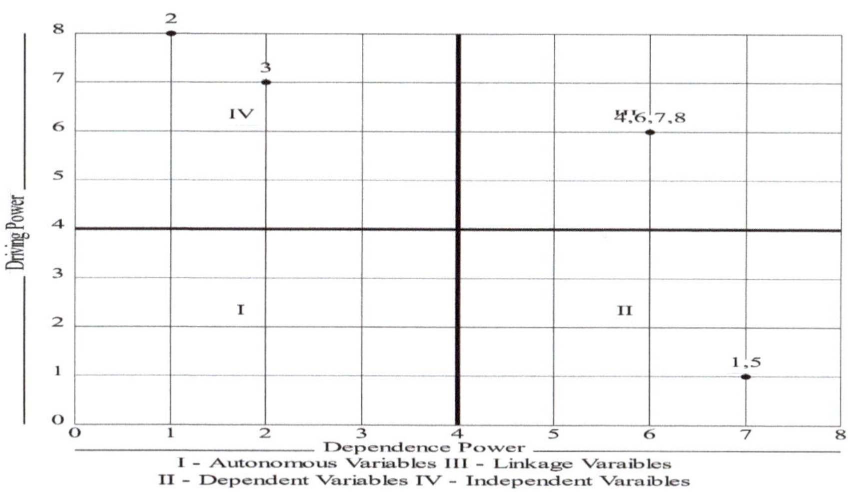

Fig. 5 Cross-impact matrix multiplication applied to classification (MICMAC) analysis of sustainability dynamic factors

9 Discussions

The study stated how different variables play diverse roles in achieving social development goals. Variables like green logistics, have restricted impact and dependence on others, while other variables for instance, labour welfare are more dependent but less influential. Crucial factors, including, innovation, technology advancement, and CSR are interrelated and need thoughtful consideration due to their variability. Green logistics occupy a dominant position over other variables but are less influenced by other factors.

10 Recommendation and Conclusion

On the basis of analysis, it is suggested that organisations should focus on crucial factors for instance, Innovation and technology Advancement, which is dominating and influential, including corporate social responsibility and transparency and trust. Line up these variables will address labour welfare and cultural resistance issues, creating a sustainable route towards social development goals. By doing this, organisations can enhance their influence on other internal and external performance and wider societal progress.

11 Limitations and Scope for the Further Research

The study narrows down and focuses on limited variables, for instance, innovation and technology advancement and Labour welfare, which may not understand the complexities of the system. Additionally, Cost Reduction and Resource Efficiency and financial incentives are crucial for further investigation.

12 Implications of the Study

By identifying crucial factors for achieving social development goals, the study suggests that policymakers and organizations incorporate and prioritize innovation, technology advancement, and CSR while designing strategies to foster sustainable growth and align with broader developmental objectives.

References

Ansu-Mensah, P. (2021). Green product awareness effect on green purchase intentions of university students': An emerging market's perspective. *Future Business Journal, 7*(1). https://doi.org/10.1186/s43093-021-00094-5

Ayu Purnamawati, I. G., Yuniarta, G. A., & Jie, F. (2023). Strengthening the role of corporate social responsibility in the dimensions of sustainable village economic development. *Heliyon, 9*(4). https://doi.org/10.1016/j.heliyon.2023.e15115

Bebbington, J., Brown, J., Frame, B., & Thomson, I. (2007). Theorizing engagement: The potential of a critical dialogic approach. *Accounting, Auditing and Accountability Journal, 20*(3), 356–381. https://doi.org/10.1108/09513570710748544

Bhosale, V. A., & Kant, R. (2016). An integrated ISM fuzzy MICMAC approach for modelling the supply chain knowledge flow enablers. *International Journal of Production Research, 54*(24), 7374–7399. https://doi.org/10.1080/00207543.2016.1189102

Bux, H., Zhang, Z., & Ahmad, N. (2020). Promoting sustainability through corporate social responsibility implementation in the manufacturing industry: An empirical analysis of barriers using the ISM-MICMAC approach. *Corporate Social Responsibility and Environmental Management, 27*(4), 1729–1748. https://doi.org/10.1002/csr.1920

Chauhan, A. S., Badhotiya, G. K., Soni, G., & Kumari, P. (2020). Investigating interdependencies of sustainable supplier selection criteria: An appraisal using ISM. *Journal of Global Operations and Strategic Sourcing, 13*(2), 195–210. https://doi.org/10.1108/JGOSS-02-2019-0017

Gardiner, R., & Hajek, P. (2020). Municipal waste generation, R&D intensity, and economic growth nexus—A case of EU regions. *Waste Management, 114*, 124–135. https://doi.org/10.1016/j.wasman.2020.06.038

Govindan, K., Shaw, M., & Majumdar, A. (2021). Social sustainability tensions in multi-tier supply chain: A systematic literature review towards conceptual framework development. *Journal of Cleaner Production, 279*. https://doi.org/10.1016/j.jclepro.2020.123075

Hina, M., Chauhan, C., Kaur, P., Kraus, S., & Dhir, A. (2022). Drivers and barriers of circular economy business models: Where we are now, and where we are heading. *Journal of Cleaner Production, 333*. https://doi.org/10.1016/j.jclepro.2021.130049

Işık, C., Ongan, S., & Islam, H. (2024). A new pathway to sustainability: Integrating economic dimension (ECON) into ESG factors as (ECON-ESG) and aligned with sustainable development goals (SDGs). *Journal of Ekonomi, 6*(1), 34–39. https://doi.org/10.58251/ekonomi.1450860

Jain, P., & Hudnurkar, D. M. (2022). Sustainable packaging in the FMCG industry. *Cleaner and Responsible Consumption, 7*. https://doi.org/10.1016/j.clrc.2022.100075

Khoshfetrat, S., Rahiminezhad Galankashi, M., & Almasi, M. (2020). Sustainable supplier selection and order allocation: A fuzzy approach. *Engineering Optimization, 52*(9), 1494–1507. https://doi.org/10.1080/0305215X.2019.1663185

Li, X., Sohail, S., Majeed, M. T., & Ahmad, W. (2021). Green logistics, economic growth, and environmental quality: Evidence from one belt and road initiative economies. *Environmental Science and Pollution Research, 28*(24), 30664–30674. https://doi.org/10.1007/s11356-021-12839-4

Lt lt S. M. (1991). Sustainable development a critical review. In: *World Development* (Vol. 19, Issue 6).

Mangla, S., Kumar, P., & Kumar Barua, M. (2014). An evaluation of attribute for improving the green supply chain performance via dematel method. *Int. J. Mech. Eng. & Rob. Res.* www.ijmerr.com

Neves, S. A., & Marques, A. C. (2022). Drivers and barriers in the transition from a linear economy to a circular economy. *Journal of Cleaner Production, 341*. https://doi.org/10.1016/j.jclepro.2022.130865

Nida, Chandra, A., & Shukla, R. (2024). ISM-fuzzy MICMAC approach for modelling the enablers of sustainability. *IIMBG Journal of Sustainable Business and Innovation*. https://doi.org/10.1108/ijsbi-11-2023-0054

Sheehy, B., & Farneti, F. (2021). Corporate social responsibility, sustainability, sustainable development and corporate sustainability: What is the difference, and does it matter? *Sustainability (Switzerland), 13*(11). https://doi.org/10.3390/su13115965

Torelli, R. (2021). Sustainability, responsibility and ethics: Different concepts for a single path. *Social Responsibility Journal, 17*(5), 719–739. https://doi.org/10.1108/SRJ-03-2020-0081

Vafadarnikjoo, A., Ahmadi, H. B., Hazen, B. T., & Liou, J. J. H. (2020). Understanding interdependencies among social sustainability evaluation criteria in an emerging economy. *Sustainability (Switzerland), 12*(5). https://doi.org/10.3390/su12051934

Yu, C., Zhao, W., & Li, M. (2019). An integrated sustainable supplier selection approach using compensatory and non-compensatory decision methods. *Kybernetes, 48*(8), 1782–1805. https://doi.org/10.1108/K-02-2018-0063

Transforming Health Care Management Through Artificial Intelligence

Reena Lenka, Rohan Deshpande, and Vivek Singh

Abstract Artificial intelligence (AI) aspires to mimic human intelligence. The rapid progress of examination techniques and the increased accessibility of medical information are changing how people view medical treatment. The current study examines how AI is used in medical services and discusses potential future uses. Information on many types of medical services may be processed using artificial intelligence (organized and unstructured). As is customary for text mining of unstructured data, well-known AI approaches also accompany techniques that can be used for organized information and data. Malignancy, nervous system science, and cardiology are important areas that use AI technology. In the three significant regions of right-on-time detection and analysis, therapy, outcome prediction, and diagnostic assessment, we further explore the nuances of AI applications in stroke. The research shows that early AI frameworks like IBM Watson can make implementing AI in real-world settings challenging.

Keywords Artificial intelligence · Vector machine · Stroke · AI frameworks

1 Introduction

Constantly monitoring a patient under specific situations involves assuring a medical professional or doctor the best possible action to commence the correct procedure or regulate the needed medicine. Cardiology, obstetrics, nervous system science, and

R. Lenka (✉) · R. Deshpande
Symbiosis Institute of Management Studies Symbiosis International (Deemed) University, Pune, India
e-mail: reena.lenka@sims.edu

R. Deshpande
e-mail: rohan.deshpande@sims.edu

V. Singh
California Intercortical University, Sioux Falls, SD, USA
e-mail: vivek.singh@caluniversity.edu

R. S. Mor et al. (eds.), *Sustainable Business Management, Innovation and Technology*, Sustainable Economy and Ecotechnology,
https://doi.org/10.1007/978-3-031-98141-8_11

brain research are just a few fields where constant remote patient surveillance is required for patient care and involves high-risk patients (Davahli, et al., 2021).

Framework and method for observing patients at high risk via remote reasoning. Most high-risk patients may be followed concurrently without involvement from the patient. The patient hears the specialist's inquiries and answers during each observing encounter. The replies from the patient are recorded at a distance focal checking station and are available for subsequent review or online (Quinn et al., 2021). To provide a patient with specified query options, Artificial intelligence. AI uses parsing to choose the patient's queries depending on their response. The screen might be made of various objects, including uterine movement strips, glaucoma circulatory strain sleeves, beat screens, electroencephalographs, etc. The optimal number of phone lines for every understanding would be four: one for the screen, one for the voice, one as a backup, and one to keep track of disappointments. The connection between a device that disperses stimulating drugs and various medical service devices, such as beat screens, thermometers, pulse screens, etc., can be seen by using double-tone grid recurrence signals (DTMF), which can transmit checked signs and other data that DEC voice, which is still under development, can understand (Wang et al., 2021) The customer must be expressly informed when to take the readings or what to do with the results.

There should be at least one source of clinical data, about one source of individual clinical information for roughly one patient, and at least one server where the clinical data and the particular clinical details are accessible to the server as part of a framework for individual clinical consideration, philosophical investigation, and conclusion(s). The server(s) might have an artificial intelligence (AI) component for analyzing each patient's clinical data and at least one interface for continuing bi-directional correspondence with at least one approved external element Haque et al. (2020)

An electronic medical services consistency assistance framework comprises a client device and a host station utilized by a healthcare professional to assist the patients regarding the method and process for filling the required drugs from the datasets and periodic check-ups. It uses packed computerized conversation to provide comprehensible and natural-sounding audio communications World Health Organization (2016). A healthcare professional uses The host station to create and maintain a database of discourse messages and related information. Information is delivered from the host station to the client device to establish a redone process. A line structure makes it easier to identify pharmaceuticals with usual dosage intervals in communications United States Government Accountability Office (2020). Allowing flexible message rescheduling is described as a boundary. When a message is prepared to play, the client device alerts the client via flexible and intelligent visible and visual signals. A satellite device may be used to produce such prompts to alert the client and advise the client device for the message (Sendak et al., 2020). To play messages, repeat messages, verify that the desired task has been done, or play messages providing help, client-worked keys are offered. The client device records each client's activities and inactions to send the information to the host station later, which collects the information into a report for medical care providers (University of Leeds (2020).

Early electronic clock and medication compliance alarms frequently required the user to configure the device using a variety of switches and fasteners. Even with the guidance of text prompts on an LCD. Some patients, particularly the elderly, find it difficult and dangerous to remember the grouping of activities and the meanings of the buttons. This is because the little fastens are difficult for them to operate Xie et al (2020).

Moreover, it is up to the patient to remember and accurately reassemble the device each time the routine changes. According to U.S. Patent No. 5,088,056, the client must operate a different mode switch depending on whether the medication is taken as directed, early, or in response to the circumstance. Hence, the customer experiences increased complexity and debilitation due to the enhanced usefulness Simonite (2020).

Moreover, complete reliance on instant messaging causes problems for older people, customers with limited competence, and persons who are visibly disabled. Such issues might result from the LCD's use of glare or the relatively small type size required for such "versatile" devices. In the unlikely event that the presentation size is increased to improve comprehension, the updated gadget gets more extensive and less valuable. The amount and type of information transmitted in text design simultaneously in a practical device is still constrained (Oktay et al., 2020).

In other instances, the visual cues go on forever without a response from the client. Many illustrations show that a persistent warning that shuts off when the client reacts might utilize too much force. Hence, under some previous craftsmanship licenses, the client can only reply to the alarm before the perceptible brief is terminated. Some do the detectable short repeatedly, on cue, or at regular intervals for a predetermined period, like 10 Alverez-Valle & Moore (2020).

Even if these and similar devices will leave an LCD message or LED pointer to remind the user to take a pill, the customer still wants additional information. For instance, LED shows how to prevent overeating by instructing a client not to take a with too little time separating it from the next portion. A client should receive more detailed instructions regarding taking prescription measurements after the schedule and doing so close to the hour of the next scheduled dose Senior et al. (2020).

Medication updates allow former employers' clients a suitable way to communicate with them. Yet, research indicates that different customers have varying levels of intellectual capacity, necessitating a range of communication strategies The AlphaFold team (2020).

The system enables a doctor-client to get updates on each sick patient by the doctor's preferred standards, subject to AI limitations Turea (2019).

A framework, technique, and computer program item that may be changed to provide a patient with a practical, complex clinical record that the patient may send is needed to ensure that access to the patient's clinical data is always available. This previously unmet need is more common in crises since it takes longer for a healthcare practitioner to assess a patient's condition and give them the care they require Reardon (2019).

2 Review Literature

This paper talks about Artificial intelligence (AI), a subfield of computer science that has the potential to dramatically change how healthcare is delivered and how medicine is performed. In this review paper, we provide a road map for developing effective, trustworthy, and safe AI systems, discuss current advancements in the application of AI in healthcare, and speculate on the potential future of AI-enhanced healthcare systems Topol (2019).

In this paper, the author talks about healthcare-related processes, from diagnosis, operations, and pathological findings of patients, which are done with the help of artificial intelligence. The critical healthcare processes can be done quickly and efficiently with few human interventions and with utmost accuracy, leading to a correct time-saving diagnosis and immediate problem addressing. This leads to more saving of life. All these benefits lead to increased dependency on Artificial intelligence in the healthcare sector (Kelly et al., 2019).

Artificial intelligence is "the theory and development of computer systems able to do activities ordinarily requiring human intellect, such as visual perception, speech recognition, decision-making, and language translation." John McCarthy was the first to introduce the phrase "artificial intelligence" to respond to Alan Turing's query from the early 1950s, "Can machines think?" The area of computer science known as artificial intelligence seeks to recreate or duplicate human intellect in machines. Artificial intelligence (AI) algorithms can make morally sound judgments besides robots' usual mechanical and planned reactions Panch et al. (2019).

Recently, more researchers have been paying more attention to artificial intelligence in the healthcare sector. Previously, research on the importance of AI in different sectors like Decision Sciences, Management, Business, and Finance has been done NHSX (2019).

The author talks about Artificial intelligence (AI) being used increasingly in healthcare due to the sector's complexity and data growth. Payers, care providers, and life sciences organizations use various AI technologies. The main application areas include suggestions for diagnosis and treatment, patient involvement and adherence, and administrative tasks. Although there are numerous cases in which AI can execute healthcare duties as well or better than humans, implementation concerns will preclude large-scale automation of healthcare professional positions for a significant duration. The use of AI in healthcare and ethical concerns are also covered Wiens et al. (2019).

Healthcare has always been swamped with a tonne of sophisticated data that is added at a rapid rate. The healthcare business generates enormous amounts of data from several sources, including hospitals and healthcare providers, medical insurance, medical gear, life sciences, and medical research. With technological innovation, there is a significant opportunity to use this data to revolutionize healthcare. When analytics, machine learning, and artificial intelligence are applied to extensive data, it is possible to identify patterns and correlations, which leads to actionable insights for bettering healthcare delivery. Although there have been several contributions to the literature on this subject, we still need a clear picture of how research and application are doing right now. This essay evaluates the material to give scholars data that enables more advancement in this field (Nachev et al., 2019).

Evaluating the literature to give the researchers information. The healthcare sector is adopting impactful technologies to enhance patient care. Artificial intelligence (AI) in healthcare and medicine has recently drawn significant attention from researchers Strodthoff & Strodthoff (2019).

Healthcare AI can replace many monitoring and tracking-related activities performed by doctors, administrators, or even patients and perform them in less time and with lower costs. This serves as the foundation for the expanded use of AI in healthcare Bellemo et al. (2019).

Furthermore, physicians benefit from AI-driven tools and applications from radiography to tumor identification. Artificial intelligence applications in healthcare uncover fresh approaches to creating new medications, one of the essential advantages of AI in healthcare Ting et al. (2019).

This paper analyzes healthcare data to improve patient care and reduce costs and clinical risks while strengthening administrative procedures inside organizations. AI in healthcare is becoming increasingly significant and has the potential to spur economic growth, alter how people work, and increase the efficacy of their efforts. As a result, integrating AI systems into healthcare can optimize available resources, enhance patient satisfaction and health, and lower patient treatment costs. In recent periods, although few studies have described the rational decision-making process for introducing new technologies into the healthcare system, the majority have concentrated on the possible advantages and challenges of integrating AI into healthcare Lashinsky (2017).

This paper talks about how healthcare, as well as the fields of medical and dental education, are being quickly transformed by artificial intelligence (AI). Healthcare and education are two fields proliferating due to AI technology breakthroughs and their incorporation into everyday work. This article aims to provide an in-depth discussion of the effects of AI in specific industries and the benefits and drawbacks of its integration. The application of AI in healthcare will be examined in the first section of the article, along with its effects on patient care, diagnosis, and treatment, as well as the advantages it offers to patients and medical professionals. The application of AI in medicine will next be discussed, along with how it affects instructional strategies and student learning, as well as the advantages and disadvantages it offers teachers and students Feeley (2017).

3 Problem Statement

AI may use specially developed algorithms to "learn" structures from a vast amount of medical information and then use the newfound understanding to support medical practice (Davahli al., 2021). Moreover, it is strengthened with learning and self-correcting abilities to advance its response-based accuracy (Quinn et al., 2021). Furthermore, AI systems can assist clinicians by providing up-to-date medical information from research books, journals, and clinical settings to illuminate optimal patient treatment. Moreover, a system using artificial intelligence (AI) can reduce therapeutic and diagnostic errors inherent in human medical practice. A schematic representation of the hybrid healthcare architecture was shown (Fig. 1). Also, an AI system extracts valuable information from a large patient population to enable real-time implications for health danger awareness and wellness (Wang et al., 2021).

Figure 2 explains the process flow of AI in the health sector. The diagram talks about the various data, which starts with a patient who goes to Private clinics and goes to the medical where their various reports like CT and Lab are collected in the reception and, with the help of the IT department, preserved in the public cloud. AI then fetches data from the public cloud and makes the information available to the patients and the doctor for further processing and treatment, which leads to smoother and error-free diagnosis and treatment.

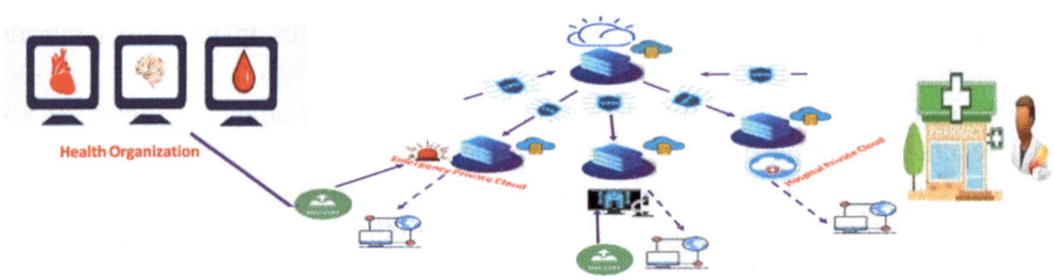

Fig. 1 Schematic representation of healthcare hybrid cloud architecture. *Source* Author

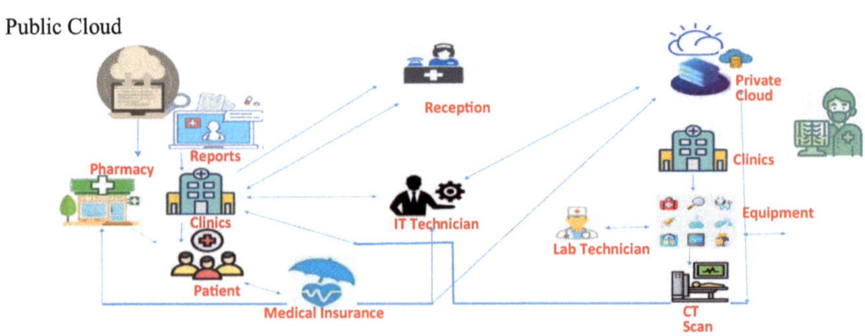

Fig. 2 General schematic view of the process flow of AI in the healthcare sector. *Source* Author

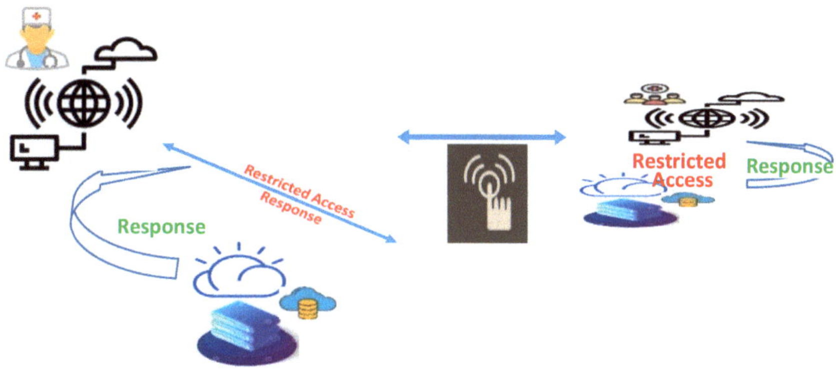

Fig. 3 Medical staff and patient remote access. *Source* Author

Figure 3 describes the medical staff and patient remote access process where patients, with the help of AI, can have restricted access to the medical staff anywhere in the globe. The patient can discuss their health-related issues by uploading their queries and reports to the cloud, which is protected, and getting suggestions from medical staff. The entire discussion is again protected from outsiders, leading to better and more timely treatment.

The most common use of classical machine learning in healthcare is accuracy medicine, which predicts which treatment procedures are likely to be successful on a patient based on several patient characteristics and their medical history.

4 Result

A training dataset for which the outcome variable (such as the onset of illness) is known is required for the extreme majority of machine learning and precision medicine demands; this is known as controlled learning (Fig. 3). The literature gathered from the study articles also revealed that the application of AI was high in the year of 2019–2020. According to the current survey, the impact of AI in the health care system is reliant mainly on recent years as compared to the prior six-year.

For a while, it was believed that lack of patient commitment and adherence were the final hurdles between poor health outcomes and effective treatment. The more frequently the usage is used and personally experiencing the resulting benefits, the more patients actively participate in their well-being and treatment. Advanced data analytics powered by AI increasingly aim to generate more confidence and engagement for its users.

5 Authors Contribution/Novelty

Unlike the previous type of brain organization, profound learning incorporates additional secret layers so that the computations can deal with complicated information with varied structures. Convolution neural organization (CNN), intermittent neural organization, profound conviction organization, and profound neural organization are the commonly used deep learning computations in clinical applications.

A modern improvement on the traditional neural organizing method is deep learning. Deep learning may be compared to a multi-layered neural structure (Fig. 4). Modern computing has advanced quickly, creating complex neural networks with many layers that were impossible for older neural networks.

Deep learning may, therefore, examine more complex non-direct instances in the data. Because of the growth in the amount and complexity of knowledge, profound learning continues to enjoy popularity (Fig. 5).

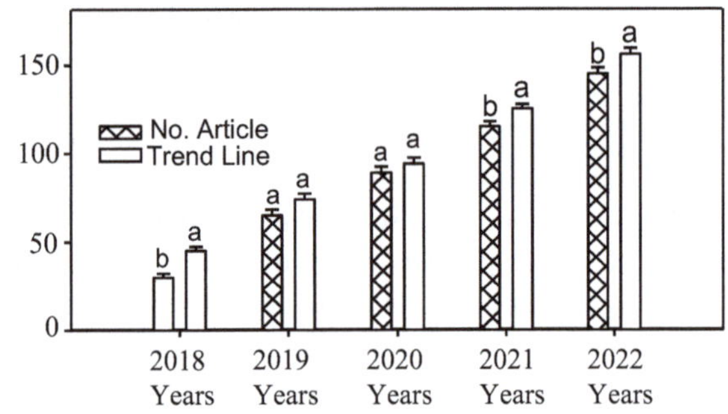

Fig. 4 Application of AI and their trend line from the year 2018–2022

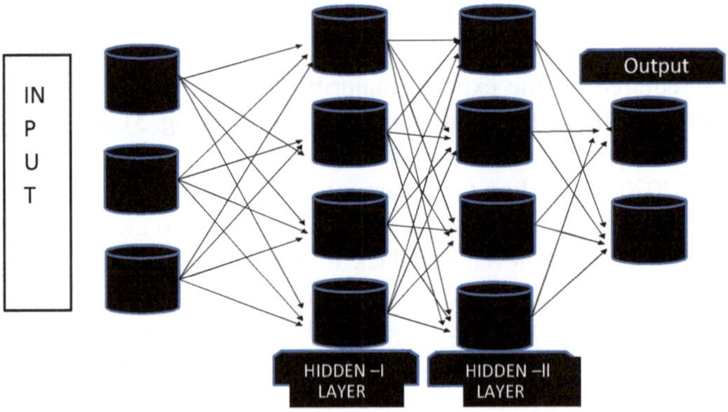

Fig. 5 Schematic view of deep learning with two hidden layers in the health care system. *Source* Author

6 Discussion and Conclusion

The present finding points to the fact that the use of Machine learning and AI is the need of the hour. Nowadays, most healthcare-related processes, right from diagnosis, operations, and pathological findings of the patients, are done with the help of Artificial Intelligence.

Critical healthcare processes can be done quickly and efficiently with few human interventions and with utmost accuracy, leading to a correct time-saving diagnosis and immediate problem addressing. This can be life-saving. All these benefits increase the reliability of Artificial intelligence in the healthcare sector.

AI in healthcare is transforming the future of healthcare in a big way. AI technologies are helping healthcare become more accurate, predictive, and time-saving. The patients are being handled with the utmost care, which leads to a more positive impact on their minds. The diagnosis and operations are becoming more accurate and less invasive, leading to quick relief from hospitals and a fast recovery rate. According to NHS, time-saving can help doctors and nurses give time to patients' care so that the patient has an overall good healthcare experience.

AI applications in healthcare are a boon, but many challenges must be addressed to make AI-enabled healthcare successful. The steps that should be implemented are as follows:

- Training for Healthcare administrators related to digital capability and proficiency in operating AI-enabled healthcare machines.
- Prior domain knowledge related to dataset operations for healthcare administrators is necessary.
- Data privacy and information security policies must be strengthened and strictly adhered to. Only a lawful use of patient records can be allowed for processing for research and development purposes.

References

Alverez-Valle, J., Moore, G. J. (2020). *Project InnerEye open-source deep learning toolkit: Democratizing medical imaging AI*. Microsoft. www.microsoft.com/en-us/research/blog/project-innereye-open-source-deep-learning-toolkit-democratizing-medical-imaging-ai

Bejnordi, B. E., Veta, M., Van Diest, P. J. et al. (2017). Diagnostic assessment of deep learning algorithms for detection of lymph node metastases in women with breast cancer. *JAMA 318*, 2199–2210.

Bellemo, V., Lim, Z. W., Lim, G. et al. (2019). Artificial intelligence using deep learning to screen for referable and vision-threatening diabetic retinopathy in Africa: a clinical validation study. *Lancet Digital Health 1*, e35

Binns, R., Gallo, V. (2019). *Trade-offs*. Information Commissioner's Office. https://ico.org.uk/about-the-ico/news-and-events/ai-blog-trade-offs

Davahli, M. R., Karwowski, W., Fiok, K., Wan, T., & Parsaei, H. R. (2021). Controlling safety of artificial intelligence-based systems in healthcare. *Symmetry 13*, 102.

Dave, M., & Patel, N. (2023). Artificial intelligence in healthcare and education. *British Dental Journal, 234*(10), 761–764.

Feeley, D. (2017). *The Triple Aim or the Quadruple Aim? Four Points to Help Set Your Strategy*. Institute for Healthcare Improvement. www.ihi.org/communities/blogs/the-triple-aim-or-the-quadruple-aim-four-points-to-help-set-your-strategy

Grazia, D., Francesca, D., Antonio, F., Matilda, S. (2023). Healthcare system: Moving forward with artificial intelligence, *Technovation 120*, 102510. ISSN 0166-4972. https://doi.org/10.1016/j.technovation.2022.102510

Haque, A., Milstein, A., Fei-Fei, L. (2020). Illuminating the dark spaces of healthcare with ambient intelligence. *Nature 585*, 193–202.

Kelly, C. J., Karthikesalingam, A., Suleyman, M., Corrado, G., King, D. (2019). Key challenges for delivering clinical impact with artificial intelligence. *BMC Medicine 17*, 195.

Lashinsky, A. (2017). *Tim cook on how Apple champions the environment, education, and health care*. Fortune.

Muoio, D. (2021). Google's next-gen Nest Hub debuts with contactless sleep monitoring and analysis features. *Mobi Health News*. www.mobihealthnews.com/news/googles-next-gen-nest-hub-debuts-contactless-sleep-monitoring-and-analysis-features

Nachev, P., Herron, D., McNally, N., Rees, G., Williams, B. (2019). Redefining the research hospital. *NPJ Digital Medicine 2*, 119.

NHSX. (2019). *Artificial intelligence: How to get it right*. NHS.

Oktay, O., Nanavati, J., Schwaighofer, A. et al. (2020). Evaluation of deep learning to augment image-guided radiotherapy for head and neck and prostate cancers. *JAMA Network Open 3*, e2027426.

Panch, T., Mattie, H., Celi, L. A. (2019). The 'inconvenient truth' about AI in healthcare. *NPJ Digital Medicine 2*, 77.

Quinn, T. P., Senadeera, M., Jacobs, S., Coghlan, S., Le, V. (2021). Trust, and medical AI: the challenges and expertise needed to overcome them. *Journal of the American Medical Informatics Association 28*, 890–894.

Raumviboonsuk, P., Krause, J., Chotcomwongse, P. et al. (2019). Deep learning versus human graders for classifying diabetic retinopathy severity in a nationwide screening program. *NPJ Digital Medicine 2*, 25.

Reardon, S. (2019). Rise of robot radiologists. *Nature 576*, S54–8.

Satya Nadella announces strategic collaboration with Novartis. (2019). *You Tube*. www.youtube.com/watch?v=wMfsQE-D2q4

Sendak, M. P., D'Arcy, J., Kashyap, S. et al. (2020). A path for translation of machine learning products into healthcare delivery. *EMJ Innov 10*, 19–00172.

Senior, A. W., Evans, R., Jumper, J. et al. (2020). Improved protein structure prediction using potentials from deep learning. *Nature 577*, 706–710.

Simonite, T. (2020). *The US government will pay doctors to use these AI algorithms*. Wired. www.wired.com/story/us-government-pay-doctors-use-ai-algorithms

Strodthoff, N., Strodthoff, C. (2019). Detecting and interpreting myocardial infarction using fully convolutional neural networks. *Physiological Measurement 40*, 015001.

The AlphaFold team. (2020). *AlphaFold: A solution to a 50-year-old grand challenge in biology*. DeepMind. https://deepmind.com/blog/article/alphafold-a-solution-to-a-50-year-old-grand-challenge-in-biology

The Royal Society. (2017). *Machine learning: the power and promise of computers that learn by example*. The Royal Society

The Health Foundation, Nuffield Trust, The King's Fund. (2018). *The health care workforce in England: make or break?* The King's Fund

Ting, D. S. W., Pasquale, L. R., Peng, L. et al. (2019). Artificial intelligence and deep learning in ophthalmology. *British Journal of Ophthalmology 103*, 167–75.

Topol, E. J. (2019). High-performance medicine: The convergence of human and artificial intelligence. *Nature Medicine 25*, 44–56.

Turea, M. (2019). *How the 'Big 4' tech companies are leading healthcare innovation.* Healthcare Weekly

United States Government Accountability Office. (2020). *Artificial intelligence in health care: Benefits and challenges of technologies to augment patient care.* GAO.

University of Leeds. (2020). *NPIC—Northern Pathology Imaging Co-operative.* University of Leeds. www.virtualpathology.leeds.ac.uk/npic

Wang, A., Nguyen, D., Sridhar, A. R. et al. (2021). Using smart speakers to monitor heart rhythms contactlessly. *Communications Biology 4*, 319.

Wiens, J., Saria, S., Sendak, M. et al. (2019). Do no harm: A roadmap for responsible machine learning for health care. *Natural Medicine 25*, 1337–1340.

World Health Organization. *Working for health and growth: Investing in the health workforce.* WHO. http://apps.who.int/iris/bitstream/10665/250047/1/9789241511308-eng.pdf. Accessed January 31, 2020.

Xie, Y., Nguyen, Q. D., Hamzah, H. et al. (2020). Artificial intelligence for teleophthalmology-based diabetic retinopathy screening in a national programme: an economic analysis modelling study. *Lancet Digit Health 2*, e240–9.

Impact of Knowledge Capital, Organisational Culture, and Innovation on Performance in the Indian IT Sector

V. Anandavel, P. Nalini, B. Elamurugan, P. Aranganathan, and G. Uppili Srinivasan

Abstract The complex linkages among Knowledge Capital (KC), Capacity for Innovation (CI), and Performance of Organisation (PO) have been extensively studied. However, conclusively linking these interactions in knowledge-driven industries like information technology (IT) is still a challenge. This work closes this gap by carefully analyzing the interactions between Culture of the Organisation (CO), KC, and PO, with CI potentially acting as a mediating factor, particularly in the framework of the changing Indian IT environment. Employing a quantitative approach, this study surveyed 486 middle-level employees of leading IT companies across key Indian cities. The research, which makes use of Smart PLS 3.2, indicates that Corporate Wealth (CW), Human Resource Capital (HRC), and CO all positively impact CI. On the other hand, CW, HRC, and CI greatly support the PO. Interestingly, PO had a non-significant impact of PO on CO. Examining the interaction in greater detail, the study shows that CI somewhat mediates the effect of PO on HRC and CW but entirely mediates the link between CO and PO. These findings offer valuable practical insights to IT leaders. Leveraging the HRC, CW, and CO through the lens of the CI can be a potent strategy for maximizing performance. Moreover, this study identifies key areas for further investigation, highlighting the evolving dynamics within the Indian IT sector. The actionable recommendations provided, along with

V. Anandavel (✉)
Department of Management, Dhanalakshmi Srinivasan College of Engineering and Technology, Mamallapuram, Chennai, Tamil Nadu, India
e-mail: anandavelmba@live.com

P. Nalini
VET Institute of Arts and science (Co-education) College, Erode, Tamil Nadu, India

B. Elamurugan
Department of Management Studies, Saveetha Institute of Medical and Technical Sciences (Deemed to be University), Chennai, India

P. Aranganathan
Hallmark Business School, Tiruchirappalli, India

G. Uppili Srinivasan
Knowledge Resource Centre (SASHE), SASTRA Deemed University, Thanjavur, Tamil Nadu, India

R. S. Mor et al. (eds.), *Sustainable Business Management, Innovation and Technology*, Sustainable Economy and Ecotechnology,
https://doi.org/10.1007/978-3-031-98141-8_12

the groundwork laid for future exploration, pave the way for a deeper understanding of this dynamic and evolving domain.

Keywords Knowledge capital · Human resource capital · Corporate wealth · Culture of the organisation · Performance of organisation · Competitive advantage

1 Introduction

In today's "Volatile, Uncertain, Complex, and Ambiguous (VUCA)" world, Knowledge Capital (KC) has become a cornerstone for societal, cultural, and economic development (Radjenovic et al., 2021). Its vital role in fostering firm success, innovation, and competitiveness has been championed by scholars, such as (Porter, 1985; Stewart, 1997), who further categorized KC into human, Organisational, and relationship capital.

In the 2023 Global Innovation Index (GII), India came in at number 40 out of 132 economies, according to the World Intellectual Property Organization (WIPO). India was placed 81st in 2015; this is an improvement. Recognizing the critical role that Capacity for Innovation (CI) plays in Performance of Organization (PO), the GII is a tool that assists governments in assessing how innovation influences social and economic changes in their nations (Edvinsson & Malone, 1997). It investigates how Culture of the Organisation (CO) can nurture innovation, as suggested by Lam et al. (2021); Saidi, 2016), and further explores the link between CI and overall performance, following (Al-Kalouti et al., 2020; Azmi & Mohd Suradi, 2019).

Despite robust growth, research on IT firms' KC is limited (Madan, 2011). This study addresses this gap by identifying CI as a mediator between KC characteristics, CO, and Performance of Organisation (PO) in the Indian IT context. Two key questions guided the investigation.

- Does PO positively correlate with Human Resource Capital, Corporate Wealth, and CO?
- Does CI mediate this relationship?

By scrutinizing these relationships, this study aims to equip managers with insights into fostering KC and nurturing the essential elements within Organisations to bridge the performance gap. Consequently, this adds significant insights to the corpus of current research in this ever-evolving subject.

2 Literature Review

The research framework draws upon Resource-Based Theory (RBT), acknowledging the vital role that resources participate in shaping a firm's offerings and performance (Grant, 1991; Wernerfelt, 1984). Effective management of both tangible and intangible resources is emphasized by RBT, such as Knowledge Capital (KC), which encompasses human, structural, relational, and technical elements—to achieve competitive advantage and enhanced performance (Asiaei & Jusoh, 2015; Bontis et al., 2000). This framework serves as the lens through which the study examines how KC attributes influence Performance of Organisation (PO).

Furthermore, recognizing the significant influence of Culture of the Organisation (CO) on business operations and innovation (Denison & Mishra, 1995; Lam et al., 2021; Saidi, 2016), the conceptual model integrates CO with HRC, CW, Capacity for Innovation (CI), and PO. This comprehensive approach allows for a deeper understanding of how KC and culture dynamically interact to affect overall performance.

2.1 *Knowledge Capital (KC) and Capacity for Innovation (CI)*

Building on the established role of Human Resource Capital (HRC) as individual capabilities, experience, and knowledge (Edvinsson & Malone, 1997), expanding research underscores its critical influence in driving innovation (Smith et al., 2012). Similarly, Structural Capital facilitating knowledge creation and utilization, known as Infrastructure of Organisation (IO) (Edvinsson & Sullivan, 1996), also contributes to innovative skill. Further, Corporate Wealth (CW) is seen as a key enabler, empowering HRC to generate greater value through effective, creative, and intelligent work (Sangkala, 2006). Informed by these insights, the following hypotheses are formulated:

Hypothesis 1: HRC positively influences an Organisation's Capacity for Innovation (CI).

Hypothesis 2: CW positively influences an Organisation's Capacity for Innovation (CI).

2.2 *Culture of the Organisation (CO) and Capacity for Innovation (CI)*

Distinct from other Knowledge Capital components, Corporate Wealth (CW) resides within the company and persists beyond employee turnover (Martín-de Castro et al., 2013; Subramaniam & Youndt, 2005). This unique characteristic aligns with its recognized impact on stimulating innovative behaviour (Çakar & Ertürk, 2010; Hartmann, 2006; Martins & Terblanche, 2003). While some debate classifies CW as part of Infrastructure of Organisation (Kaul & Singh, 2019), its prevailing influence on Capacity for Innovation (CI) prompts the following hypothesis:

Hypothesis 3: CO has a favourable impact on CI.

2.3 *Capacity for Innovation (CI) and Performance of Organisation (PO)*

Mounting evidence (Beltramino et al., 2020; Saunila & Ukko, 2012) points to a strong link between Performance of Organisation (PO) and Capacity for Innovation (CI). The development of new offerings is fuelled by innovation, which is essential to an Organisation's existence and expansion (Yang, 2012). Hence, we formulate the following hypothesis:

Hypothesis 4: CI positively impacts PO.

2.4 *Knowledge Capital (KC) and Performance of Organisation (PO)*

Building on (Maran et al., 2009) definition of Human Resource Capital (HRC) development, which strengthens skills, potential, and engagement, according to this study, HRC has a favourable impact on Performance of Organisation (PO). Evidence from Bontis et al. (2000); Rahim et al., 2011) further supports this connection, highlighting the impact of Infrastructure of Organisation (IO) on PO across diverse sectors. The following hypothesis was developed in light of these revelations:

Hypothesis 5: HRC positively influences PO.

Hypothesis 6: CO positively influences PO.

2.5 *Culture of the Organisation (CO) and Performance of Organisation (PO)*

While data points to a connection between Culture of the Organisation (CO) and performance (PO) (Acar, 2012; Gorondutse & Hilman, 2019; Polychroniou & Trivellas, 2017), conflicting findings (Leithy, 2017; Rashid & Shah, 2016) necessitate further investigation. Recognizing this complexity, the following hypothesis is proposed:

Hypothesis 7: CO may positively influence PO.

2.6 *Mediation Impact of Capacity for Innovation (CI) on Human Resource Capital (HRC), Infrastructure of Organisation (IO), Culture of the Organisation (CO), and Performance of Organisation (PO)*

Capacity for Innovation (CI) bridges Knowledge Capital (KC), Human Resource Capital (HRC), structural (IO), social capital (SOC), and Performance of Organisation (PO) (Citrasari et al., 2022). Its mediating role amplifies the impact of KC on Culture of the Organisation (CO), potentially boosting competitive advantage (Ali & Anwar, 2021) and firm performance.

Extensive research has confirmed HRC's significant influence of HRC on PO through CI (Dhar et al., 2019; Xiaobo & Sivalogathasan, 2013). Similarly, IO's positive association between IO and PO via CI is well-established (Abdirahman & Tarique, 2020). The advantageous of the CO in PO are also underscored (Al-Kalouti et al., 2020; O'Cass & Sok, 2013; Oura et al., 2016).

However, this relationship in the Indian IT sector requires further exploration. Notably, research on CO's impact of CO on PO, mediated by CI, remains limited. Thus, in light of the literature review, we put out the following hypothesis:

Hypothesis 8: HRC's impact on PO is mediated by CI.

Hypothesis 9: CO's influence on PO is mediated by CI.

2.7 *Conceptual Model*

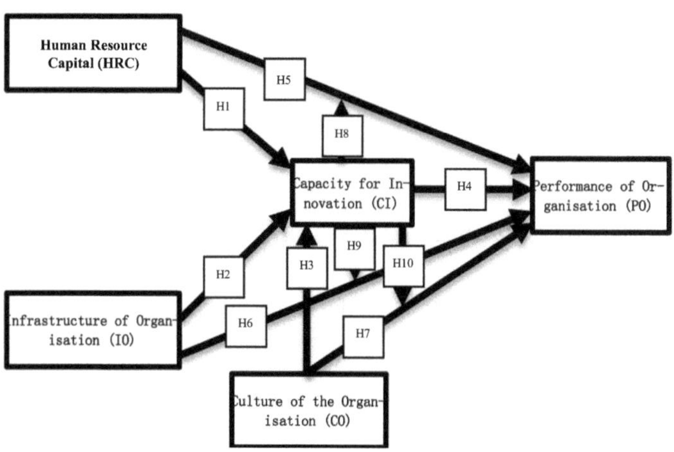

3 Methods

3.1 Sampling Procedure and Data Collection

The study used a survey methodology to gather information from middle-level staff members in leading Indian IT companies (Tata Consultancy Services (TCS), Infosys Limited, HCLTech, Wipro Limited, and Tech Mahindra) across key cities, such as Bangalore, Hyderabad, and Chennai. Targeting a population of more than 50,000 professionals with 3–10 years of experience, the estimated sample size for this large group (> 50,000) was 384. Convenience sampling was utilized to get 486 useable replies from the 560 questionnaires that were distributed in order to ease practical data collection. The subsequent data analysis was conducted using SPSS 25.0.

3.2 Sampling Procedure and Data Collection

The structured questionnaire was meticulously designed, leveraging the instruments established in prior research. A Likert scale with five consistent points (1 being strong disagreement and 5 being strong agreement) facilitated comprehensive data collection across all the constructs. Human Resource Capital (HRC) and Infrastructure of Organisation (IO) were assessed using validated scales from Subramaniam & Youndt (2005); Bontis et al., 2000). Performance of Organisation (PO), the dependent variable, was quantified using the scales established by (White et al. 1998); Sousa, 2004). Leveraging (Maher, 2014) work, Questions were designed to assess the influence of the Culture of the Organisation (CO) on Capacity for Innovation (CI). Finally, scales provide a robust measurement of the mediating variable, CI (Sousa, 2004; White et al., 1998).

4 Results

4.1 Analysis of Reliability and Validity

Using the procedures given in Table 1, the constructs' validity and reliability were carefully evaluated. If the inclusion of a component considerably increased composite reliability or average variance extracted (AVE) above acceptable limits, components with weak outer loadings (0.40–0.70) were taken into consideration for removal. Despite exceeding the recommended limits, removing certain items further improved the CR and AVE. After eliminating 11 indicators, confidence interval analysis confirmed no outer loadings near zero.

Table 1 Fornell–Larcker criterion of correlations

Construct	Innovation capability	Human capital	Organizational capital	Organizational culture	Organizational performance
Capacity for innovation (CI)	0.890				
Human resource capital (HRC)	0.820	0.780			
Corporate wealth (CW)	0.250	0.710	0.650		
Culture of the organisation (CO)	0.680	0.670	0.040	0.610	
Performance of organisation (PO)	0.720	0.750	0.100	0.620	0.740

Table 2 Heterotrait–Monotrait ratio of correlations

Construct	Innovation capability	Human capital	Organizational capital	Organizational culture	Organizational performance
Capacity for innovation	–				
Human resource capital	0.85	–			
Corporate wealth	0.26	0.14	–		
Culture of the organisation	0.81	0.8	0.12	–	
Performance of organisation	0.87	0.88	0.13	0.74	–

Dependability was evaluated using Cronbach's alpha, rho a, and composite reliability, all exceeding the minimal cutoff point of 0.70. Rho a, surpassing 0.70 and falling between Cronbach's alpha and CR, further highlighted robust reliability. Convergent validity was demonstrated by an AVE greater than 0.5. As shown in Tables 2 and 3, additional analyses utilizing the heterotrait-monotrait ratio of correlations and the square root of the AVE validated the discriminant validity.

4.2 *Structural Model*

This study used R2, Q2, and significance tests to investigate the proposed routes in the structural model. R2 values of 0.610 for Capacity for Innovation and 0.630 for Performance of Organisation, exceeding the 0.1 threshold recommended by, confirmed the

Table 3 Testing direct relationships

Hypotheses	Path Coefficients'	SD	't' value (Bootstrap)	'p' values	BI	
					2.50%	97.50%
H1: HRC → CI	0.47	0.06	7.85	0.00	0.12	0.35
H2: CW → CI	0.21	0.04	5.30	0.00	0.05	0.14
H3: CO → CI	0.37	0.05	6.40	0.00	0.14	0.24
H4: CI → PO	0.36	0.07	5.50	0.00	0.11	0.24
H5: HRC → PO	0.45	0.07	6.50	0.00	0.27	0.32
H6: CW → PO	0.14	0.04	4.08	0.00	0.04	0.10
H7: CO → PO	0.09	0.08	1.20	0.23	0.21	– 0.06
R^2 Capacity for innovation = 0.610						
R^2 Performance of organisation = 0.630						
Q^2 Capacity for innovation = 0.335						
Q^2 Performance of organisation = 0.340						

Abbreviations: BI—bias corrected confidence interval; HRC—"Human Resource Capital"; CW—"Corporate Wealth"; CO—"Culture of the Organisation"; CI—"Capacity for Innovation"; PO—"Performance of Organisation".

model's ability to predict future outcomes. All Q2 values were greater than 0, with Q2 Capacity for Innovation at 0.335 and Q2 Performance of Organisation at 0.340, further reinforcing the model's significant predictive capacity.

Hypothesis testing revealed support for five out of the seven proposed relationships:

H1: HRC → CI was accepted using a path coefficient (β) of 0.47, t Value (Bootstrap) of 7.85, p-value of 0.00, and Bootstrap Interval (BI) from 0.12 to 0.35.

H2: CW → CI was accepted using a path coefficient (β) of 0.21, t Value (Bootstrap) of 5.30, p-value of 0.00, and BI from 0.05 to 0.14.

H3: CO → CI was accepted using a path coefficient (β) of 0.37, t Value (Bootstrap) of 6.40, p-value of 0.00, and BI from 0.14 to 0.24.

H4: CI → PO was accepted using a path coefficient (β) of 0.36, t Value (Bootstrap) of 5.50, p-value of 0.00, and BI from 0.11 to 0.24.

H5: HRC → PO was accepted using a path coefficient (β) of 0.45, t Value (Bootstrap) of 6.50, p-value of 0.00, and BI from 0.27 to 0.32.

However, two hypotheses were not supported:

H6: CW → PO was rejected, indicating no significant effect using a path coefficient (β) of 0.14, t Value (Bootstrap) of 4.08, p-value of 0.00, and BI from 0.04 to 0.10.

H7: CO → PO was also rejected, indicating no significant effect using a path coefficient (β) of 0.09, t Value (Bootstrap) of 1.20, p-value of 0.23, and BI from 0.21 to − 0.06.

4.3 Mediation Analysis

Table 5 reveals intriguing findings regarding mediation analysis. Both direct and indirect pathways were significant for HRC → CI → PO (H8) and CW → CI → PO (H9), indicating partial mediation. This suggests that HRC and CW influence PO partially through CI, but also exert independent effects. H10, however, presents a distinct picture. The non-significant direct link between CO and PO, coupled with a significant indirect effect, underscores the full mediation. Thus, CO influences PO solely through CI, acting as a catalyst for Organisational growth via Capacity for Innovation (Table 4).

Table 4 Mediation Analysis (Indirect Effect)

Hypotheses	Coefficients'	SD	't' Value (Bootstrap)	'p' values	BI	
					2.50%	97.50%
H8: Human Resource Capital (HRC) → Capacity for Innovation (CI) → Performance of Organisation (PO)	0.163	0.041	4.256	0.000	0.095	0.211
H9: Corporate Wealth (CW) → Capacity for Innovation (CI) → Performance of Organisation (PO)	0.157	0.018	4.321	0.000	0.042	0.149
H10: Culture of the Organisation (CO) → Capacity for Innovation (CI) → Performance of Organisation (PO)	0.123	0.031	4.611	0.000	0.076	0.113

Table 5 Mediation analysis (direct effect)

Human resource capital (HRC) → Performance of organisation (PO)	0.427	0.000
Corporate wealth (CW) → Performance of organisation (PO)	0.135	0.000
Culture of the organisation (CO) → Performance of organisation (PO)	0.087	0.249

5 Discussion

This study offers a comprehensive analysis of the interrelationships among Human Resource Capital (HRC), Corporate Wealth (CW), Capacity for Innovation (CI), and Performance of the Organisation (PO), particularly within the Indian IT sector. The findings reaffirm that HRC plays a critical role in enhancing CI, underscoring the importance of skilled and creative human capital in fostering innovation. This aligns with prior literature (Çakar & Ertürk, 2010; Saunila & Ukko, 2012), which emphasizes that innovative individuals are vital for navigating organizational challenges. The study also highlights a significant correlation between CW and CI, consistent with (Yang, 2012), suggesting that corporate financial strength and resource availability directly influence an organization's ability to innovate. Furthermore, a meaningful relationship between the Culture of the Organisation (CO) and PO confirms earlier research (Bontis et al., 2000; Rahim et al., 2011), indicating that organizational culture is a key driver of overall effectiveness. One of the most notable contributions of this research is the mediation analysis, which reveals that CI fully mediates the effects of HRC and CO on PO. This finding positions CI as the critical mechanism through which human capital and organizational culture translate into enhanced performance. Interestingly, CW's impact on PO is only partially mediated by CI, suggesting a more nuanced and complex pathway.

These findings collectively validate the hypothesized linkages between HRC, CW, CO, and CI and their combined influence on PO, thus enriching the theoretical understanding of Knowledge Capital (KC) in organizational contexts. The validated model confirms the strategic significance of aligning KC development with innovation capacity to drive superior outcomes. Practically, the study emphasizes that IT executives should prioritize the development of HRC, enhance CW, and cultivate a conducive CO to foster innovation. Such investments are not merely HR or cultural decisions but strategic imperatives that significantly impact PO. The research also serves as a valuable resource for policymakers in the IT sector, encouraging them to design frameworks that facilitate KC expansion and innovation-driven performance.

In summary, this study bridges critical knowledge gaps by exploring the mediating role of CI and its interaction with key organizational resources. It delivers significant theoretical and managerial insights into how innovation capacity can be leveraged to enhance organizational performance in a dynamic, knowledge-intensive industry. The strategic alignment of HRC, CW, CO, and CI offer a holistic pathway for sustainable organizational growth and competitive advantage.

6 Conclusion

This study deconstructs the intricate dimensions of Human Resource Capital (HRC), Infrastructure of Organisation (IO), Culture of the Organisation (CO), and Capacity for Innovation (CI) to assess their impacts on Information Technology (IT) firms.

Recognizing the knowledge-intensive nature and IP-driven competition in IT, this study aimed to provide insights into Performance of Organisation determinants. The findings reveal a healthy association between firm performance and KC, CO, and CI, enriching the existing knowledge about their interplay within the IT sector. However, this study acknowledges the complexity of success in IT and advocates for a broader understanding. Therefore, it recommends integrating factors such as Relational Capital, Environmental Uncertainty, and Knowledge Sharing in future IT and technology-based research. This aligns with the dynamic nature of the IT landscape, in which evolving challenges and opportunities demand a broader scope of investigation. Future study can obtain a more comprehensive understanding of the complex dynamics impacting an Organisation's performance by integrating these components, keeping up with the dynamic factors that determine success in the IT industry.

References

Abdirahman, M. D., & Tarique, R. (2020). Impact of structural capital and innovation capability on firm performance: Case study of Pharma industry in Karachi-Pakistan. *The Strategic Journal of Business & Change Management, 7*(1), 736–748.

Acar, A. Z. (2012). Organizational culture, leadership styles and organizational commitment in Turkish logistics industry. *Procedia—Social and Behavioral Sciences, 58*, 217–226. https://doi.org/10.1016/j.sbspro.2012.09.995

Ali, B. J., & Anwar, G. (2021). Business strategy: The influence of strategic competitiveness on competitive advantage. *International Journal of Electrical, Electronics and Computers, 6*(2).

Al-Kalouti, J., Kumar, V., Kumar, N., Garza-Reyes, J. A., Upadhyay, A., & Zwiegelaar, J. B. (2020). Investigating innovation capability and organizational performance in service firms. *Strategic Change, 29*(1), 103–113.

Asiaei, K., & Jusoh, R. (2015). A multidimensional view of intellectual capital: The impact on organizational performance. *Management Decision, 53*(3), 668–697.

Azmi, R., & Mohd Suradi, N. (2019). Innovation capability and organization performance in Malaysian public sector: A conceptual framework. *AIP Conference Proceedings, 2111*, Article 020016. https://doi.org/10.1063/1.5111223

Beltramino, N. S., García-Perez-de-Lema, D., & Valdez-Juárez, L. E. (2020). The structural capital, the innovation and the performance of the industrial SMES. *Journal of Intellectual Capital, 21*(6), 913–945. https://doi.org/10.1108/JIC-01-2019-0020

Bontis, N., Keow, W. C. C., & Richardson, S. (2000). Intellectual capital and business performance in Malaysian industries. *Journal of Intellectual Capital, 1*(1), 85–100.

Çakar, N. D., & Ertürk, A. (2010). Comparing innovation capability of small and medium-sized enterprises: Examining the effects of organizational culture and empowerment. *Journal of Small Business Management, 48*(3), 325–359.

Citrasari, P. D., Ratnawati, K., & Hadiwidjojo, D. (2022). Influence of intellectual capital on operational performance through innovation capability in Indonesia's defence industry. *Media Ekonomi dan Manajemen 37*, 252–272.

Denison, D. R., & Mishra, A. K. (1995). Towards a theory of organizational culture and effectiveness. *Organization Science, 6*, 204–223.

Dhar, B. K., Mutalib, M., Sobhani, F. A., & Business, F. (2019). Effect of innovation capability on human capital and organizational performance. *International Journal of Advanced Science and Technology, 29*(4), 7074–7087.

Edvinsson, L., & Malone, M. S. (1997). *Intellectual capital: Realizing your company's true value by finding its hidden brainpower*. Harper Business.

Edvinsson, L., & Sullivan, P. (1996). Developing a model for managing intellectual capital. *European Management Journal, 14*, 356–364. https://doi.org/10.1016/0263-2373(96)00022-9

Gorondutse, A. H., & Hilman, H. (2019). Does organizational culture matter in the relationship between trust and SMEs performance. *Management Decision 57*, 1638–1658.

Grant, R. M. (1991). The resource-based theory of competitive advantage: Implications for strategy formulation. *California Management Review, 33*(3), 114–135.

Hartmann, A. (2006). The role of organizational culture in motivating innovative behavior in construction firms. *Construction Innovation, 6*(3), 159–172.

Kaul, A., & Singh, A. (2019). Structural Capital: A study of select organizations. *Amity Business Review, 19*(2), 2018.

Lam, L., Nguyen, P., Le, N., & Tran, K. (2021). The relation among organizational culture, knowledge management, and innovation capability: Its implication for open innovation. *Journal of Open Innovation: Technology, Market, and Complexity, 7*(1), 66. https://doi.org/10.3390/joitmc7010066

Leithy, W. (2017). Organizational culture and organizational performance. *International Journal of Economics & Management Sciences, 06*, 1–10. https://doi.org/10.4172/2162-6359.1000442

Madan, A. (2011). Indian higher education and the need for critical knowledges. *Contemporary Education Dialogue, 8*(2), 161–182.

Maher, L. (2014). Building a culture for innovation: A leadership challenge. *World Hospitals and Health Services: The Official Journal of the International Hospital Federation, 50*, 4–6.

Maran, M., Arokiasamy, L., & Ismail, M. (2009). Human capital development and its impact on firm performance: evidence from developmental economics. *Journal of International Social Research, 2*.

Martín-de Castro, G., Delgado-Verde, M., Navas-López, J. E., & Cruz-González, J. (2013). The moderating role of innovation culture in the relationship between knowledge assets and product innovation. *Technological Forecasting and Social Change, 80*(2), 351–363.

Martins, E., & Terblanche, F. (2003). Building organisational culture that stimulates creativity and innovation. *European Journal of Innovation Management, 6*(1), 64–74.

O'Cass, A., & Sok, P. (2013). The role of intellectual resources, product innovation capability, reputational resources and marketing capability combinations in firm growth. *International Small Business Journal, 32*(8), 996–1018. https://doi.org/10.1177/0266242613480225

Oura, M. M., Zilber, S. N., & Lopes, E. L. (2016). Innovation capacity, international experience and export performance of SMEs in Brazil. *International Business Review, 25*(4), 921–932. https://doi.org/10.1016/j.ibusrev.2015.12.002

Polychroniou, P., & Trivellas, P. (2017). The impact of strong and balanced organizational cultures on firm performance. *International Journal of Quality and Service Sciences, 10*(1), 16–35. https://doi.org/10.1108/IJQSS-09-2016-0065

Porter, M. E. (1985). *Competitive advantage* (p. 557). New York: Free Press.

Radjenovic, T., Zivkovic, S., & Todorović, M. (2021). The role of intellectual capital in the economic growth: the case of the EU countries. In *Economic and Social Development, 73rd International Scientific Conference on Economic and Social Development—"Sustainable Tourism in Post-pandemic World"*. Dubrovnik

Rahim, A., Atan, R., & Kamaluddin, A. (2011). Intellectual capital reporting in Malaysian technology industry. *Asian Journal of Accounting and Governance, 2*. Retrieved from http://www.ukm.my/ajag/vol2ch5.pdf

Rashid, H., & Shah, A. B. A. (2016). Impact of organizational culture on organizational performance: Evidence from education sector. *Journal of Business Management and Economic Studies, 1*(1), 16–30.

Saidi, A. A. (2016). The impact of organizational culture on innovation capability of SMEs—case study of SMEs in Alimosho and Ojo local government area of Lagos State, Nigeria. *International Journal of Economics, Commerce and Management, United Kingdom, IV*(9), 158–181.

Sangkala. (2006). Intellectual capital management: Strategi Baru Membangun Daya Saing. Yogyakarta: YAPENSI.

Saunila, M., & Ukko, J. (2012). A conceptual framework for the measurement of innovation capability and its effects. *Baltic Journal of Management, 7*(4), 355–375. https://doi.org/10.1108/17465261211272139

Smith, M. B., Bryan, L. K., & Vodanovich, S. J. (2012). The counter-intuitive effects of flow on positive leadership and employee attitudes: Incorporating positive psychology into the management of organizations. *The Psychologist Manager Journal, 15*(3), 174–198.

Sousa, C. M. (2004). Export performance measurement: An evaluation of the empirical research in the literature. *Academy of Marketing Science Review, 9*, 1–22.

Stewart, T. A. (1997). *Intellectual capital: The new wealth of organizations.* Doubleday/Currency.

Subramaniam, M., & Youndt, M. A. (2005). The influence of intellectual capital on the types of innovative capabilities. *Academy of Management Journal, 48*, 450–463. https://doi.org/10.5465/AMJ.2005.17407911

Wernerfelt, B. (1984). A resource-based view of the firm. *Strategic Management Journal, 5*(2), 171–180.

White, S. D., Griffith, D. A., & Ryans, J. K. (1998). Measuring export performance in service industries. *International Marketing Review, 15*(3), 188–204.

Wu, X., & Sivalogathasan, V. (2013). Intellectual capital for innovation capability: A conceptual model for innovation. *International Journal of Trade, Economics and Finance, 4*(3). https://doi.org/10.7763/IJTEF.2013.V4.274.

Yang, J. (2012). Innovation capability and corporate growth: An empirical investigation in China. *Journal of Engineering and Technology Management—JET-M, 29*(1), 34–46. https://doi.org/10.1016/j.jengtecman.2011.09.004

E-Tailing Business Performance in the Age of Artificial Intelligence

Sarita Chaudhary, Mamta Gupta, and Gaurav Bhardwaj

Abstract Revenues from online retail business continue to climb up around the globe and the application of Artificial Intelligence has dramatically changed the way, the business performs activities, which were earlier being performed by human beings. Artificial intelligence helps online retailers follow business trends, changing customer needs in the market. The company can gather a wide range of information, assess customers, and then adequately respond to their requirements and habits by providing quality services. With the help of AI, online retailers are also getting numerous benefits such as operational efficiency, cost reduction, consumer satisfaction, competitive advantage and many more. The objective of this study is to find out the driving factors that influence the online retail business using AI technologies and their impact on the overall performance of businesses. Primary data were collected from a self-administered questionnaire through a mail survey of 125 small and medium sized online retail firms, which eventually led to 114 valid questionnaires. Analytical results indicate that the surveyed factors, operational efficiency, consumer experience, digital access and cost effectiveness positively affect the growth and performance of online retailers in India.

Keywords Artificial intelligence · Consumer behaviour · Retailing · Chatbots · Business performance

S. Chaudhary · M. Gupta · G. Bhardwaj (✉)
Department of Management, Maharaja Surajmal Institute, New Delhi, India
e-mail: gauravbhardwaj@msijanakpuri.com

S. Chaudhary
e-mail: saritachaudhary@msijanakpuri.com

M. Gupta
e-mail: mamtadelin@gmail.com

R. S. Mor et al. (eds.), *Sustainable Business Management, Innovation and Technology*,
Sustainable Economy and Ecotechnology,
https://doi.org/10.1007/978-3-031-98141-8_13

1 Introduction

Technology is still changing the game in retail, which can be advantageous to both consumers and merchants. For instance, technology enables customers to make better judgments, receive more advantageous offers that are more specifically suited to them, and receive speedier service (Gupta et al., 2022) Due to technologically developed efficiencies, it also helps retailers reach the right people at a cheaper cost. Artificial intelligence is growing in popularity because of transforming the retail landscape with the advancement of information and communication technology, which can benefit both customers and businesses (Haenlein & Kaplan 2019).

Consumer needs continue to guide their purchasing decisions in the quickly changing retail environment. Most consuming decisions are made by consumers, but emerging technologies like the Internet of Things, robotics, subscription models, and big data/predictive analytics indicate that shopping is about to take a quantum leap into an uncharted territories (Grewal et al., 2009), as a result, there is a strong need to comprehend key retailing sectors where innovations are upending the status quo to better understand how the industry will develop going forward. Customers may make better decisions, obtain more beneficial offers that are better suited to them, and receive faster service thanks to the use of artificial intelligence (Nagy & Hajdú 2021). In the current, competitive market, the retail sector must now more than ever provide customers with cutting-edge, high-quality services. Additionally, AI technologies support retailers by opening a wide range of opportunities for innovation processes that lead to superior customer service, effective supply chain management, optimized digital marketing, risk management, increased sales, quick expansion, competitive advantage, and effective operations, all of which increase the profitability of those companies (Guha et al., 2021). Most industries are adopting the still-expanding digital retail marketing. From fashion and cosmetics to food, the need for the adoption of a digital alternative is spreading throughout the retail sectors. Mobile technology also enables retailers to show pertinent offers that take locational information (such as time of day, weather, and location) into account (Bradlow et al., 2017). This is done by using location-based applications like Google maps. Feel-good merchandise, for instance, can sell well in bad weather.

Most industries are embracing digital retail marketing, which is still expanding quickly. The retail sector will need to continuously adapt and evolve to thrive against the escalating rivalry brought on by further technological advancements like the advent of artificial intelligence, blockchains, virtual reality, and more (Inman & Nikolova 2017). As evidenced by the introduction of self-checkout lanes, chatbots powered by artificial intelligence, and so much more, this shift has already started. Because frequent, digital contacts between businesses and customers produce greater datasets, consumer industries like retail and high tech typically perceive more potential in AI applications in marketing and sales.

2 Impact of Artificial Intelligence on Online Retailers

It is difficult to deal with each customer individually because online retailing draws in a lot of customers and they all have distinct options. According to industry history and current trends, AI technology is used to detect organizations' needs and preferences as well as those of their customers (Kiruthika & Khaddaj 2017). AI has helped generate a variety of fresh concepts that aid in understanding consumers' purchasing habits. AI technologies have several benefits, from planning tasks to administrative tasks along the value chain (Siau & Yang 2017). Numerous data-generating gadgets have made it essential for firms to adopt new tactics and approaches that will give managers access to correct information quickly. Using program-based algorithms, AI may automate procedures, extract knowledge from historical data, and present users with consumer or market insights that aid managers in making decisions. AI can analyse a consumer's habits, purchases, likes, and dislikes. Customer Relationship Management (CRM) software, chatbots, and other technologies have significantly benefited from AI. Chatbots are examples of artificial intelligence (AI) assistants that automatically react to straightforward customer enquiries with e-voice instructions and offer ideas to consumers using natural language processing. Based on machine learning algorithms, chatbots can act just like people. Chatbots have a significant impact on customer service at online retailers since they are available around-the-clock, assist customers in finding products, have access to the most recent supply chain data, compare different products, and show customers how to make payments (Kietzmann et al., 2018). Additionally, the addition of this software can lower labour costs since employees won't have to spend time fielding repetitive questions; instead, the chatbot can handle simple interactions, and for more complicated ones, it will help the customer get in touch with the appropriate personnel. These benefits improve the experiences of customers, as well as increase the value of online businesses.

The retail sector in India is one of the largest and fastest-growing sectors, accounting for 10% of all employment and more than 13% of the nation's GDP. India presents a wide range of opportunities for e-retailers as a developing market. Additionally, during the epidemic, online buying, which offered greater flexibility, interactivity, customisation, and little risk, attained astounding popularity throughout the nation (Bradlow et al. 2017). In the current era of digitalisation, as companies are transforming, AI is becoming the need of online retail firms and customers, AI has a positive impact on the business world, it has changed the traditional mood of doing business, transformed it digitally, and improved a lot in the context of ease of doing business, etc. (Inman & Nikolova 2017). Online retail marketing encompasses the complete process of making an online purchase. This process, which includes looking through many possibilities, choosing, and paying for the product, typically starts when a person searches for a product they would like to buy and finishes when they receive it. Now that individualized marketing efforts are available, e-commerce companies may engage with customers more successfully. Based on past interactions with customers, present requirements, and particular tasks, a personalized intelligent automation system can react. Businesses can increase the effectiveness of their site search engines by using AI-powered search engines, commonly referred to as insight engines. The management of inventories has been significantly impacted by predictive analysis of data science services. Inventory management is a problem when it comes to keeping shelves stocked, inventory current, and everything accessible in the supply chain. Retailers can now foresee their company's real-time inventory needs, online retailing has already been significantly impacted by AI and machine learning technology, and this impact will only grow. The number of businesses utilizing AI has increased by more than 25% in recent years. Nevertheless, due to a lack of knowledge about the supplemental resources required to improve the performance of these systems, many organizations have not yet realized the full potential of AI. Organizations struggle most with inefficiency when it comes to integrating the system with the data and using the pertinent data to train AI. Additionally, enterprises must realize that the entire process must be altered to benefit from automation; simply automating specific phases won't be sufficient to make these systems profitable for the company.

3 Problem Statement

It was evident from a survey of the literature that no detailed research had been done on how AI affected the success of online retail businesses in India. Therefore, an effort has been made to cover this issue in the suggested research. It has aided us in developing strategies to meet the latest AI-related difficulties and take full benefit of AI in online retail business.

4 Research Objectives

To identify the factors affecting business performance of online retailers using Artificial Intelligence tools.

To measure the impact of influencing factors on the business performance of online retailers.

5 Methodology

Various small, medium, and large online retail businesses have been targeted in this paper in order to understand the impact of the use of AI in their businesses. The main objective of this paper was to identify the main factors that are important to online retailers using AI technologies for the competitive advantage and business growth. The paper employed both primary and secondary data. A systematic questionnaire was used to gather the main data. While secondary data was gathered through publications like journals, books, textbooks, online articles, etc. A structured questionnaire was created, and 125 questionnaires were distributed to senior executives, managers, and business owners of various online retailers using AI techniques. Of those, 114 questionnaires were chosen for finalization and were fully completed by the respondents.

6 Measurements of the Constructs

The study explores four reflective dimensions that influence the online retail business entities to make utilization of AI technologies in their business operations, namely operational efficiency, customer experience, cost effectiveness and digital accessibility. These constructs were explored with the help of Exploratory Factor analysis method. Following is the explanation of four dimensions taken in the present study in relation to "Adoption of AI".

Operational efficiency: Operational excellence in relation to the organization's consumer-serving functions is essential for every business to succeed (Golcher-Barguil et al., 2019). Artificial intelligence transforms operational excellence by utilizing various core functionalities to enhance organization operations. AI-enabled tools are assisting in scaling retail business growth handling a large volume of concurrent online customer interactions, which improves customer satisfaction by enabling the company to offer quicker response times, higher-quality products, customer services, predicting customer demand and sales forecasting, and reduced costs (Dogru & Keskin 2020).

Customer Experience: Enhancing customer experience has become a difficult challenge for a corporation due to the constant rise in client expectations. Companies should consider more than just delivering material at the right moment through the right channel if they want to provide a competitive advantage in customer service (Awamleh & Ertugan 2021). Ameen et al. (2021) described how artificial intelligence (AI) has changed the customer experience by enabling companies to provide relevant and tailored interactions. Customers can interact with AI-powered chatbots and virtual assistants in real-time while receiving personalized recommendations, round-the-clock help, sentiment analysis, multilingual support, and evaluation of client data including purchase history and browsing habits (Mishra & Mukherjee 2019). The retailer's new reality is AI-powered customer support (Ahmad et al., 2017).

Digital accessibility: AI has advanced significantly over the past two years to assist merchants with producing original content for their websites and items. Beginning with brand-new photos for your website and blogs, as well as original digital marketing commercials. Most firms utilize advertisements to connect with their target market. Using information about a customer's browsing habits, demographics, and interests, AI systems can tailor content (Davenport et al., 2020). AI can assist marketers in tailoring content, offers, and advertisements. This boosts user experience and aids companies in developing more conversion-boosting, focused marketing efforts. AI systems can spot patterns in user data and foretell which consumers will interact with advertisements (Grandinetti 2020) This gives organizations more in-depth understanding of consumer behaviour, personalizes communications, forecasts future behaviour, and optimizes campaigns in real-time. AI is also utilized to come up with concepts or sources of inspiration for marketing content (Wirth 2018). In a matter of seconds, AI algorithms can generate post ideas, hashtag suggestions, and image captions. But in addition to coming up with these concepts from new, they can also evaluate your current material, identify its essential points, and then use that information to create interesting social media postings (Kietzmann et al., 2018). By producing high-quality, focused content that is suited to user intent, utilizing an AI tool for SEO can help your website rank better. Without having to spend a lot of time manually optimizing each piece, it also makes the process of creating optimized content simpler and faster.

Cost effectiveness: In transactional data, machine learning algorithms can spot odd patterns or departures from typical behaviour. The algorithms gain the ability to distinguish between genuine transactions and suspicious activities that can be signs of fraud by "training" on historical data, (Awamleh & Ertugan 2021). By automating repetitive procedures, businesses can save time and resources while allowing people to concentrate on more strategic goals. By lowering human error and bias, raising accuracy and efficiency, and offering insights and advice, AI can assist firms in improving their budgeting and sales forecasts (Smith 2018). To boost sales and profit margins, companies can optimize their resource allocation, cash flow management, pricing, and marketing tactics by using accurate and trustworthy sales predictions. Businesses who employ AI effectively gain insights about market dynamics, consumer trends, and competitor actions that may be leveraged to adjust offers, operations, and costs.

Application of Artificial Intelligence in E-Tailing Business Performance

7 Data Analysis and Results

A comprehensive review of literature was conducted to search for quality measures for the research constructs in the proposed model (Fig. 1). An online self-administered questionnaire was used to collect data with a total of valid 114 responses. Exploratory Factor Analysis (EFA) using a principal component analysis method was conducted on 21 scale items of the variables. One item was dropped because of high cross loadings. Factor Analysis technique extracted four factors namely "Operational Efficiency", "Customer Experience", "Cost Effectiveness", "Digital Accessibility" and accounted for approximately 66.05% of the total variance.

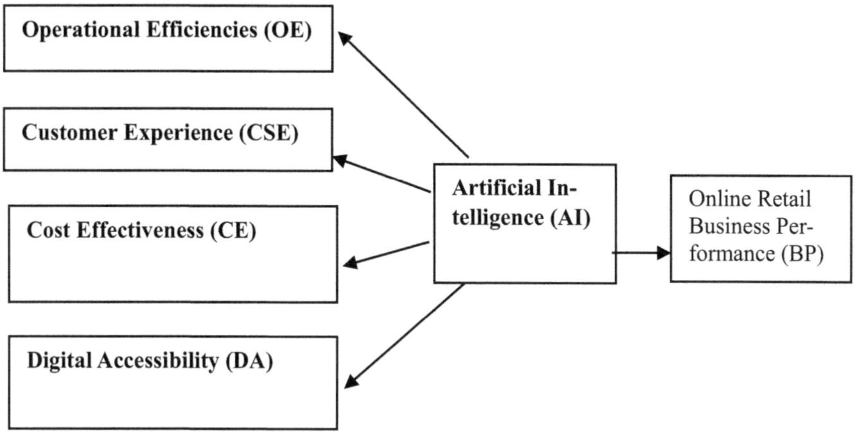

Fig. 1 Conceptual framework

8 Assessment of Measurement Model

Regression Analysis technique was applied to test hypothesis that there is no significant impact of Application of Artificial Intelligence (AI) on the Organization Performance of the Retail Businesses. Various factors related to AI have been identified by applying Exploratory Factor Analysis. The assessment of the measurement model was done through Item Loadings and Reliability of the constructs. The item loadings above 0.70 indicate that all the items provide sufficient load in their underlying constructs. Cronbach alpha value of each factor greater than 0.70 indicate that the instrument is reliable, thus confirming internal consistency (Tables 1 and 2).

AI and Business Performance have significant correlation coefficient value 0.712. It is evident that AI and Business Performance have positive relationship with each other. Also there exists a significant relationship between all the constructs at 0.01 level.

9 Regression Analysis

To validate that there is significant impact of application of AI on Business Performance, a regression model was developed. The tables below show the outcome of the regression analysis. The application of Artificial Intelligence is significantly influencing the Firms' performance as depicted by p value (< 0.05) (Table 3).

The table above ascertains the association of applications of Artificial Intelligence and Business Performance in the structural model. The beta and t- values analysed that the hypothesis is rejected, "There is no significant influence of application of Artificial Intelligence on the performance of businesses", thus proving that there is

Table 1 Exploratory factor analysis (EFA)

Factor extracted	Item scale	EFA loadings	Cronbach (Reliability)
Operational efficiencies (OE)	OE1: Quick and effective information processing	0.76	0.81
	OE2. Reduce human errors	0.79	
	OE3. Scalability in business and spots irregularities	0.82	
	OE4. Time saving, threat hunting and security tasks	0.79	
	OE5: Inventory Optimization	0.87	
Customer experience (CSE)	CSE1. Personalized recommendation of products	0.87	0.85
	CSE2. Instant support 24/7	0.87	
	CSE3. Anticipate customers' needs and potential issues	0.68	
	CSE4. Sentiment analysis	0.78	
	CSE5. Multilingual support	0.89	
Cost effectiveness (CE)	CE1. Detect fraud & secure payments	0.81	0.81
	CE2. Decision making through real-time analytics)	0.82	
	CE3. Price Optimization	0.75	
	CE4. Predicting consumer buying behaviour and limiting customer returns	0.75	
	CE5. Elimination of repetitive tasks	0.66	
Digital accessibility (DA)	DA1. Content Marketing assistance	0.74	0.82
	DA2. Automated e-mails	0.81	
	DA3. Optimization of Google ads	0.81	
	DA4. Enhance visibility of website	0.84	
	DA5. Social media marketing assistance	0.73	
Business performance (BP)	BP1. Competitive advantage	0.83	0.80
	BP2. Customer retention	0.76	

Table 2 Correlation analysis

Construct	AI	Business performance
AI	1	0.712**
Sig (2-tailed)		0.000
Business performance	0.712**	1
Sig (2-tailed)	0.000	

**Correlation is significant at 0.01 level (2-tailed)

Table 3 Regression coefficients

Model		Unstandardized coefficients		Standardized coefficients	t	Sig.
		B	Std. error	Beta		
1	(Constant)	0.454	0.014		2.357	0.02
	OE	0.026	0.065	0.31	39.9	0.00
	CSE	0.048	0.060	0.21	79.4	0.00
	CE	0.011	0.063	0.17	17.2	0.00
	DA	0.033	0.069	0.27	48.5	0.00

a. Dependent Variable: BP

significant positive impact of the application of AI on the performance of the retail businesses.

Thereby proving that there is a significant impact of "operational efficiency", "customer experience" "cost effectiveness" and "digital accessibility due to AI on the performance of online retail businesses. Standardised regression weights of factors- "operational efficiency", "customer experience" "cost effectiveness" and "digital accessibility came as positive, revealing a positive impact on retail business performance.

10 Discussion and Implications

The amount of data that is now readily available on the internet has grown exponentially over the last several years, which has sped up the development of artificial intelligence (AI) because it has access to a wealth of knowledge that it can draw on.

The global epidemic has also fundamentally altered how companies communicate with each other and with customers. Businesses can now incorporate cutting-edge tools into their routine decision-making processes, such as forecasting future sales, inventory tracking to bring operational and cost efficiency, and enhancing customer experience with the company and its service thanks to increased digitalization and interest in online business.

This research has explored certain important factors that contribute to the business performance of online retailers namely operational efficiency, customer experience, cost effectiveness and digital accessibility, focusing on the surveyed four factors, the study was able to achieve its second goal of examining the impact of applying AI techniques on the business performance of online retailers. The study looked at how the system's efficiency, speed, and accuracy improve when AI-based software is used for online retail shop functions like billing, sales returns, etc. Utilizing AI approaches, businesses have reduced costs by eliminating obsolete stock and bad loans, among other things. The use of AI benefits consumers as well, resulting in customer happiness and preserving client loyalty. Application of AI technology boosts corporate performance significantly by increasing productivity, accuracy, and lowering operating expenses. It is also revealed during the study that artificial intelligence has been a blessing for start-ups because they heavily rely on its methodologies to investigate new business opportunities. In India, where unemployment is already a problem for the nation, there is a concern that AI may worsen the situation. The employment of artificial intelligence in business has a negative impact on the unskilled labour, but it benefits intellectual people who are successful in acquiring the necessary abilities to use AI technique. Businesses need to get on board with AI sooner rather than later, or else they'll have to suffer the consequences. It's a fact of life in the tech world, and it all depends on how long it takes them to get on board. The longer they wait, the more they'll have to endure.

11 Suggestion

AI have a big impact on online retailers, giving them the chance to increase productivity, enhance customer experience, cost efficiency and digital accessibility, however, there are drawbacks associated with the adoption of these technologies, including employment displacement, privacy and security concerns, and bias, high Infrastructure cost. Maintenance of the infrastructure required for AI techniques demands for higher maintenance as well as additional resources such as training to the employees for managing these techniques. Before using AI technologies, retailers must carefully consider these variables and make sure they have the resources and experience to use these technologies efficiently. AI tools will continue to influence the future of online retailing as they become more widely available and reasonably priced, reshaping the sector in ways we can't yet envision.

Online business must make a complete shift from manual to automated processes if they are to fully capitalize on the potential of AI technologies. This involves educating employees about these technologies and establishing a human–machine relationship in which humans act as moderators and intervene or modify the AI-provided solutions in accordance with all legal and ethical considerations as well as the needs of the organization.

References

Ahmad, A., Rahman, O., & Khan, M. N. (2017). Exploring the role of website quality and hedonism in the formation of e-satisfaction and e-loyalty: Evidence from internet users in India. *Journal of Research in Interactive Marketing 11*(3), 246–267. https://doi.org/10.1108/JRIM-04-2017-0022

Ameen, N., Tarhini, A., Reppel, A., & Anand, A. (2021). Customer experiences in the age of artificial intelligence. In *Computers in Human Behaviour*. https://doi.org/10.1016/j.chb.2020.106548

Awamleh, F., & Ertugan, A. (2021). The relationship between information technology capabilities, organizational intelligence, and competitive advantage. *SAGE Open, 11*(2), 21582440211015200. https://doi.org/10.1177/21582440211015201
Bradlow, E. T., Gangwar, M., Kopalle, P., & Voleti, S. (2017). The role of big data and predictive analytics in retailing. *Journal of Retailing, 93*, 76–96.
Davenport, T., Guha, A., Grewal, D., & Bressgot, T. (2020). How artificial Intelligence will change the future of marketing. *Journal of the Academy of Marketing Science 48*, 24–42. https://doi.org/10.1007/s11747-019-00696-0
Dogru, A. K., & Keskin, B. B. (2020). AI in operations management: Applications, challenges, and opportunities. *Journal of Data Information and Management, 2*, 1–8. https://doi.org/10.1007/s42488-020-00023-1
Gólcher-Barguil, L. A., Nadeem, S. P., & Garza-Reyes, J. A. (2019). Measuring operational excellence: An operational excellence profitability (OEP) approach. *Production Planning & Control, 30*, 682–698. https://doi.org/10.1080/09537287.2019.1580784
Grandinetti, R. (2020). How artificial intelligence can change the core of marketing theory. *Innovative Marketing 16*(2), 91–103. https://doi.org/10.21511/im.16(2).2020.08
Grewal, D., Levy, M., & Kumar, V. (2009). Customer experience management: An organizing framework. *Journal of Retailing, 85*(1), 1–14.
Guha, A., Grewal, D., Kopalle, P. K., Haenlein, M., Schneider, M. J., Jung, H., & Hawkins, G. (2021). How artificial intelligence will affect the future of retailing. *Journal of Retailing, 97*(1), 28–41.
Gupta, S., Modgil, S., Bhattacharyya, S., et al. (2022). Artificial intelligence for decision support systems in the field of operations research: Review and future scope of research. *Annals of Operations Research, 308*, 215–274.
Haenlein, M., & Kaplan, A. (2019). A brief history of artificial intelligence: On the past, present, and future of artificial intelligence. *California Management Review, 61*(4), 5–14.
Inman, J. J., & Nikolova, H. (2017). Shopper-facing retail technology: A retailer adoption decision framework incorporating shopper attitudes and privacy concerns. *Journal of Retailing, 93*, 7–28.
Kietzmann, J., Paschen, J., & Treen, E. (2018). Artificial intelligence in advertising, how marketers can leverage artificial intelligence along the consumer journey. *Journal of Advertising Research, 58*(3), 263–267. https://doi.org/10.2501/JAR-2018-035
Kiruthika, J., & Khaddaj, S. (2017). Impact and challenges of using of virtual reality & artificial intelligence in businesses. In *Computing and Applications to Business, Engineering and Science* (pp.165–168). https://doi.org/10.1109/DCABES.2017.43
Mishra, N., & Mukherjee, S. (2019). Effect of artificial intelligence on customer relationship management of Amazon in Bangalore. *International Journal of Management 10*(4), 168–172. https://doi.org/10.34218/IJM.10.4.2019.016
Nagy, S., & Hajdú, N. (2021). Consumer acceptance of the use of artificial intelligence in online shopping: evidence from Hungary. *Amfiteatru Economic Journal, 23*(56).
Siau, K. L., Yang, Y. (2017). Impact of artificial intelligence, robotics, and machine learning on sales and marketing. In: *Twelve Annual Midwest Association for Information Systems Conference* (pp. 18–19).
Smith., S. (2018). *Juniper research: retailer spending on AI to grow nearly fourfold, reaching $7.3 billion by 2022*, Businesswire, viewed 28 April 2023.
Wirth, N. (2018). Hello marketing, what can artificial intelligence help you with? *International Journal of Market Research, 60*(5), 435–438.

Environmental Sustainability Practices of Select HEIs in India: An Empirical Study

Jyoti Paul, Rajesh Kumar Abhay, Dishant Bhola, Saumy Omar, and Shreya

Abstract Environmental deterioration is increasing with rapid growth and development in many sectors of society. Higher Education Institutions (HEIs) are important in promoting sustainability by educating students and implementing sustainable practices. The purpose of this paper is to evaluate the contribution of select Indian Higher Educational Institutions to sustainability, identify areas where they are lagging, and recommend ways for improvement. A novel environment sustainability index consisting of 39 parameters for the Indian context was developed after reading the extant literature and sustainability reports of HEIs in India. This index was developed to gauge the sustainability practices of select HEIs. For the study, three Indian Institutes of Management (IIMs), viz IIM Bangalore, IIM Udaipur, and IIM Lucknow, were selected as the sustainability reports of these three institutions were available. The reports for the years 2023–24 were taken for content analysis. Further, the reports and websites of the institutions were extensively analyzed to code the required data. The study found studies in the international context related to environmental sustainability in HEIs, but very few studies were found in the Indian context. Further, in Indian studies, the environmental index for HEIs could not be traced. This lack of an Index to Indian HEIs was the major impetus behind this study. A thorough investigation of such practices through a comprehensive index will provide HEIs with their strengths and weaknesses to improve in the future. The study recommends that HEIs can significantly improve their sustainability impact by adopting a holistic approach.

Keywords Higher education institutions · Environment · Sustainability index · India

---

J. Paul · D. Bhola · S. Omar
Department of Commerce, Dyal Singh College, University of Delhi, New Delhi, India
e-mail: jyotipaul@dsc.du.ac.in

R. K. Abhay (✉)
Department of Geography, Dyal Singh College, University of Delhi, New Delhi, India
e-mail: rkabhay@dsc.du.ac.in

Shreya
Department of Commerce, Dyal Singh Evening College, University of Delhi, New Delhi, India

R. S. Mor et al. (eds.), *Sustainable Business Management, Innovation and Technology*, Sustainable Economy and Ecotechnology,
https://doi.org/10.1007/978-3-031-98141-8_14

1 Introduction

Over the past decades, environmental degradation has accelerated alongside economic growth, with global challenges such as climate change, global warming, and pollution. Research by WHO has estimated that around 7 million people every year die from exposure to fine particles in polluted air. India being a vast nation, the demand for housing, food, water, and energy places a heavy burden on the environment. According to the Global Forest Watch monitoring project, India has lost 2.33 million ha of tree cover since 2000, with 144,000 ha lost in 2023 (PTI, 2024). Energy consumption has more than doubled in India since 2000. India's energy sector depends heavily on coal as a fuel for its thermal power plants, which is a major source of the emission of greenhouse gases. Sustainable development is a means of combating these challenges by directing our efforts toward a future characterized by balanced economic growth and environmental conservation. These days, a lot of organizations, including companies, healthcare facilities, and educational institutions, are adopting and incorporating sustainable practices into their daily operations as they recognize the importance of sustainability.

Higher Educational Institutions (HEIs) play a very crucial role in sustainability since they are important sources of knowledge that influence future leaders, policies, and practices that support sustainable development across a range of industries through research, innovation, and education. Confronting issues and suggesting strategies to incorporate sustainability into all facets of HEIs can facilitate its promotion and equip students for a sustainable future. Many national and international universities are actively engaged in various sustainable programs and practices. For instance, according to the Oxford University Report on Environment Sustainability (2023–24), it is involved in activities like the installation of solar photovoltaic panels around the university and the recycling of non-hazardous waste or using it to generate energy.

Keeping the above discussion into consideration, the objective of this paper is to identify and analyze the engagement of HEIs in India in environment sustainability and investigate their participation in environmental initiatives. The paper also extensively reviews literature and finds very few studies in the Indian HEIs context related to environment sustainability. The paper tries to bridge the gap and develops an Environment index particularly for Indian HEIs to gauge their initiatives.

2 Review of Literature

The exploration of sustainability practices in higher education institutions (HEIs) has revealed various strategies, achievements, and challenges across global contexts. Sustainability in HEIs is a multi-faceted concept, encompassing governance, social equity, environmental responsibility, and educational innovation. Several studies have explored these dimensions in depth. Larrán Jorge et al. (2015) examined corporate governance and institutional frameworks for sustainability across Spanish HEIs. Bormann & Nikel (2017) emphasized the significance of stakeholder engagement and adaptive governance structures in German universities. De la Poza et al. (2021), discussed the integration of Sustainable Development Goals (SDGs) into institutional policies. Caputo et al. (2021) explored the importance of transparency in sustainability reporting and the integration of social and environmental objectives. Environmental management is another critical aspect of sustainability. Nagi (2023) recommended that HEIs pursue ISO-14001[1] certification to enhance sustainability practices. Filho et al. (2021) highlighted gaps in green purchasing and sustainable catering in Latin American universities, underscoring the need for more comprehensive frameworks.

Sustainability initiatives within HEIs differ based on regional and institutional contexts. At the global level, universities in Australia, as explored by Sen et al. (2022), have achieved milestones in carbon neutrality, with institutions like the University of Tasmania and Charles Sturt University being certified as carbon–neutral. Tabucanon et al. (2021) analyzed UI Green Metric indicators to assess sustainability in large and small HEIs. Xiong & Mok (2020) examined the sustainability efforts of the Hong Kong Sustainable Campus Consortium, and have revealed that these universities have standardized sustainability practices reporting and data collection. Studies from India by Kansal & Venkatesh (2020) have focused on water sustainability alone and concluded that though institutions have water-related courses, the distribution is skewed, and they focus more on the technological aspects of water sustainability. Singh et al. (2023) have done a systematic literature review and found a list of Sustainability Assessment Tools list which have qualitative factors and quantitative factors of assessment used in different tools. Another study by Yadav & Yadav (2017) investigated business management institutions in India and their approach to incorporating sustainability into curricula.

Research on sustainability in HEIs has evolved significantly over the past decades. Earlier studies focused primarily on environmental responsibility and campus operations, but more recent work has adopted a more integrated view of sustainability. Rohrich & Takahashi (2019) provided a historical overview of environmental sustainability research in Brazilian HEIs. However, Ragazzi & Ghidini (2017) critically analysed its limitations, such as lack of scoring thresholds and high sensitivity to ranking methodologies. Most recently, Waheed et al. (2011) introduced the DPSEEA Sustainability Index Model (D-SiM), which assesses sustainability through the Driving Force-Pressure-State-Exposure-Effect-Action framework. This paper, based on gaps in literature, has focused primarily on environment aspects of sustainability in HEIs in India by developing an extensive scorecard particularly for India. The scorecard emphasizes parameters included in disclosures mandated by the National Assessment and Accreditation Council and those outlined in sustainability reports published by higher education institutions, aimed at identifying strengths and weaknesses and recommending areas for improvement in the environmental sustainability of such institutions.

Table 1 Comparative description of the selected IIMs

Particulars	IIMB	IIMU	IIML
Type of institution	Business school & Institute of national importance	Business school & Institute of national importance	Business school & Institute of national importance
Year of establishment	1973 (3rd IIM in India)	2011 (13th IIM)	1984(4th IIM)
Total campus area	Over 100 acres	300 acres (Approx.)	185 acres
Location	Urban	Lakeside	City
Focus	Diverse programs	Sustainability focus	Diverse programs
Programmes offered on sustainability	Strategy and the sustainable enterprise	Sustainable supply chain management	Postgraduate programme in sustainable management
Waterbody area	No lakes or ponds on the campus; focus on rainwater harvesting	IIMU's campus is designed around water conservation and incorporates several lakes	No lakes or ponds on the campus

Source Compiled by Authors from various sources

3 Materials and Methods

3.1 Selected Institutions for Study

This paper analyses various sustainability practices that different Indian Institutes of Management adopt. For this purpose, three Indian Institutes of Management (IIM) were selected, viz. IIMB, IIML and IIMU. All three IIMs are business schools and institutes of national importance. Other details are mentioned in Table 1.

In summary, all three IIMs are business schools with of national importance, they differ in terms of their location, focus, and programs offered on sustainability. These management institutions were chosen for the purpose of the study as Management institutes have the responsibility to equip future managers with skills to have a sustainable future. Further, as the reports of only these three IIMs were available for the year 2023–24 in public domain so the sample was restricted to these three for uniformity.

3.2 Data Source

The data for the research is collected through their website and the reports that they published, including the sustainability report and carbon footprint report. The data for the evaluation of sustainable practices in IIMB was collected from their Sustainability Report (IIMB, 2025) and Carbon Footprint Report (IIMB, 2023); data for IIML was collected from their Sustainability Report (TERI, 2024), and for IIMU from their Sustainability Policy and Report (Xavier & Upendra, 2023).

3.3 Research Methods

The method of content and statistical analysis is used to analyze the data based on the environmental index developed in the Indian context consisting of 39 parameters, which were further divided into the sub-indices as waste, water, green campus, energy, environmental governance, and environment research (Table 2). IIMs have been rated from 0 to 2 based on their involvement in a particular parameter. They have been rated '0' if no information is mentioned about the parameter, '1' if partial information has been mentioned, and '2' if full information with supporting evidence is available.

4 Results and Discussions

4.1 Environmental Parameters and IIMs Performance

Based on the selected indicators, IIMB performed the best with a score of 48, which is 61.54% of the total score, indicating a strong adherence to sustainability practices across various parameters (Table 3). IIML has a balanced performance, achieving 50% of the possible score. IIMU scored the lowest, obtaining 42.31%, which highlights areas for significant improvement. All institutions show moderate variability in their scores, with IIML having the highest (0.90) and IIMU the lowest (0.83). This indicates that scores across parameters are somewhat consistent, with IIML showing slightly more fluctuation. IIMB is slightly left-skewed (-0.47), indicating more high scores. IIML is slightly right-skewed (0.32), indicating more low scores. IIMU is Symmetrical (0.00), indicating balanced scores. All institutions have negative kurtosis, indicating flatter distributions with fewer extreme values.

The mean value of the category indicates the average performance, with IIMB leading with a 1.23 mean value. The middle score is 1.0 for all, indicating a balanced distribution around the central value. IIMB has a mode of 2.0, showing frequent high scores. IIML and IIMU have 0.0 as a mode, indicating several parameters scored zero. Overall, the Range value is 2: All institutions have a range of 2, showing the difference between the highest and lowest scores. This indicates that all institutions have parameters that score both the lowest (0) and highest (2) possible values.

Table 2 Selected parameters under different sub-indices of environment

S. No.	Parameters (Weight)	Category
1	Policy for water recycling (0, 1, 2)	Water
2	Waste-water treatment on-campus (0, 1, 2)	Water
3	Tracking of water consumption and water reuse (0, 1, 2)	Water
4	Rainwater harvesting (0, 1, 2)	Water
5	Water conservation activities on-campus (0, 1, 2)	Water
6	Water conservation activities off-campus (0, 1, 2)	Water
7	Groundwater recharge (0, 1, 2)	Water
8	Recycling of waste (0, 1, 2)	Waste
9	Minimize food waste by composting (0, 1, 2)	Waste
10	Provision of colour-coded dustbins to aid waste segregation (0, 1, 2)	Waste
11	Biogas plant installed on-campus (0, 1, 2)	Waste
12	Solid waste generation and segregation (annually and monthly) (0, 1, 2)	Waste
13	Use of natural fertilizers made onsite through composting from wet waste (0, 1, 2)	Waste
14	Paper consumption report (0, 1, 2)	Waste
15	Policy for minimizing use of single-use plastic (0, 1, 2)	Waste
16	Reviews to identify areas where energy waste is highest (0, 1, 2)	Energy
17	Active use of energy harvesting concepts (0, 1, 2)	Energy
18	Sensor-based energy conservation (0, 1, 2)	Energy
19	Net zero energy building (0, 1, 2)	Energy
20	Use of renewable sources of energy (0, 1, 2)	Energy
21	Solar water heaters (0, 1, 2)	Energy
22	Process for carbon management and reducing carbon dioxide emissions (0, 1, 2)	Energy
23	Reduce air, water, sound pollution (0, 1, 2)	Green infrastructure
24	Green office established (0, 1, 2)	Green infrastructure
25	Green policy and projection (0, 1, 2)	Green infrastructure
26	Use of digital display instead of banners and flexes (0, 1, 2)	Green infrastructure
27	Sustainable infrastructure (0, 1, 2)	Green infrastructure
28	Local environmental project and partnership (0, 1, 2)	Green infrastructure
29	Biodiversity in building plan (0, 1, 2)	Green infrastructure

(continued)

Table 2 (continued)

S. No.	Parameters (Weight)	Category
30	Plantation activities (0,1,2)	Green infrastructure
31	Use of bicycle/battery-powered vehicles on-campus (0,1,2)	Green infrastructure
32	Climate action plan (0, 1, 2)	Environmental governance
33	Master plan compliance with environmental norms (0, 1, 2)	Environmental governance
34	Environmental monitoring and impact assessment (0, 1, 2)	Environmental governance
35	Publications by faculty related to dimensions of sustainability in journals (0, 1, 2)	Environment research
36	Students doing research in sustainability (0, 1, 2)	Environment research
37	Sustainability start-ups (0, 1, 2)	Environment research
38	Compliance with standard international norms (0, 1, 2)	Environment research
39	Sustainability club (0, 1, 2)	Environment research

Source Adapted and modified from Paul et al. (2025)

Table 3 Environment parameters (Overall) descriptive statistics

Institution	Mean	Median	Mode	Total score (Out of 78)	Percentage score (%)	Count	Standard deviation	Skewness	Kurtosis
IIMB	1.23	1	2	48	61.54	39	0.84	– 0.47	– 1.43
IIML	0.85	1	0	33	50.00	39	0.9	0.32	– 1.74
IIMU	1	1	0	39	42.31	39	0.83	0	– 1.54

Source Calculated by Authors

4.2 *Performance on Water-Related Parameters*

The performance of selected IIMs on water management indicators has been shown in Table 4. It shows that IIML excels in water management with an overall score of 12 out of 14. It highlights that IIML has strong policies for water recycling, wastewater treatment, tracking water usage, and groundwater recharge. IIMB performs well as a total of 10 out of 14 though it lacks tracking water consumption and water recycling.

IIMU falls behind with a score of 5 and lacks in water recycling and wastewater treatment.

Water recycling is the process of treating wastewater and reusing it for various purposes. It can be observed that IIML is actively engaged in water recycling, while only partial information is available for IIMB. There is no evidence of involvement by IIMU in this area.

IIML and IIMB have wastewater treatment on-campus, but IIMU has no information about wastewater treatment. Wastewater treatment plants installed in IIML's campuses have impressive capacities. IIMB and IIML are actively engaged, whereas IIMU provides partial information about this parameter. When we talk about IIML, the Lucknow campus has a higher water consumption rate (175,000 L/day) compared to NOIDA (25,650 L/day). All the water conservation activities have resulted in not just increasing the water table of IIMB but also covering 20% of our domestic water needs from water generated from within the campus. Rainwater harvesting is prevalent in both IIMU and IIMB. Information about IIML is not clear. On the IIMU campus, rooftop rainwater is harvested at 6 locations and stored in 6 sumps after basic filtration. There are 5 sumps of 50,000 L capacity and 1 sum of 1 lakh liters capacity. All three IIMs have water conservation activities on-campus. IIMU has implemented several water-saving measures, such as the installation of low-flow faucets and toilets and the harvesting of rainwater. These measures have helped to reduce the institute's water consumption and conserve natural resources. There is no information available regarding water conservation activities off-campus in any of the IIMs. IIMB and IIML are actively engaged, whereas IIMU provides no information about this parameter. IIML has taken a proactive approach. The Lucknow campus boasts 10 artificial recharge pits with a combined capacity of 700 m^3 and 2 natural pits capable of storing a staggering 27,320 m^3 of water. At IIMB, several recharge wells were dug across the campus. IIMB has dug 57 large holes to collect rainwater.

Table 4 Performance of IIMs on water management indicators

Water management indicators	IIMB	IIMU	IIML
Policy for water recycling (0, 1, 2)	1	0	2
Wastewater treatment on-campus (0, 1, 2)	2	0	2
Tracking of water consumption (0, 1, 2)	1	2	2
Rainwater harvesting (0, 1, 2)	2	1	2
Water conservation (on-campus) (0, 1, 2)	2	2	2
Water conservation (off-campus) (0, 1, 2)	0	0	0
Groundwater recharge (0, 1, 2)	2	0	2
Total Score (Out of 14)	10	5	12
Percentage Score	71.4	35.7	85.7

Source Authors

4.3 Waste Management Practices and Sub-indices

Waste management practices in HEIs is an important aspect of environmental sustainability. Under the same category, IIMB achieves a score of 13 out of 16, demonstrating comprehensive waste management practices such as waste recycling, composting, and waste segregation. IIMU scores 3 (Table 5). It shows some involvement in waste segregation but also lack in other key areas like composting and biogas plants. IIML is at par with IIMU, indicating very limited engagement in waste management apart from waste segregation and provision of paper consumption report.

It can be observed that IIMB is actively engaged in waste recycling. They use natural fertilizers, made onsite through the composting of wet waste. IIMU provides partial information on the recycling of waste, stating in brief that waste generated on-campus is recycled. There is no evidence of involvement by IIML in this area. IIMB is on the path to becoming a zero-waste campus, they are partner with their service providers on the reduction of waste, but there is no such information regarding these activities is provided by IIMU and IIML. IIMB has provisions for color-coded dustbins on its campus to aid waste segregation, whereas IIMU and IIML do not have such provisions. Food waste is converted into compost and biogas by microbial cultures in a biogas plant designed and installed by IIMB alumni, on-campus. There is no information available regarding the presence of such a plant at IIMU or IIML. At IIMB, waste collection and sorting are managed on-campus with the assistance of experts from an NGO called SAHAS. Additionally, dry waste is meticulously sorted and sent to authorized recyclers. IIML is actively engaged in the segregation of waste. IIMU provides brief information about this parameter. IIMB uses natural fertilizers. There is no information provided by IIMU and IIML regarding such practices on their campus. IIML mentions briefly about their paper consumption, while there is no such information provided by IIMB and IIMU. IIMB states that they use digital displays,

Table 5 Performance on waste parameters

Waste management indicators	IIMB	IIMU	IIML
Recycling of waste (0, 1, 2)	2	1	0
Minimizing food waste by composting (0, 1, 2)	2	0	0
Color-coded dustbins for waste segregation (0, 1, 2)	2	0	0
Biogas plant installed on-campus (0, 1, 2)	2	0	0
Solid waste generation and segregation (0, 1, 2)	2	2	2
Use natural fertilizers made onsite through composting from wet waste (0, 1, 2)	2	0	0
Paper consumption report (0, 1, 2)	0	0	1
Policy for minimizing use of single-use plastic (0, 1, 2)	1	0	0
Total score (Out of 16)	13	3	3
Percentage score	81.3	18.8	18.8

Source Authors

instead of banners and flexes. Additionally, they also state that their practices to reduce waste involves avoiding the use of plastic. IIMU and IIML mentions no such information about their consumption and use of plastic.

4.4 Energy Management and Conservation

HEIs in India are improving their efforts to combat with energy crisis. It has been found that IIMB performs best in energy management with a total score of 8 out of 14 excelling in areas like renewable energy, solar water heaters, and carbon management (Table 6). However, it falls short in advanced energy practices like energy harvesting and Net Zero buildings. IIMU and IIML tie with a 50% score demonstrating efforts in renewable energy but missing out on more advanced practices.

IIMB has implemented comprehensive carbon management strategies, including various energy-saving measures and initiatives to reduce its carbon footprint. Similarly, IIML actively works on reducing CO_2 emissions. On the other hand, IIMU has partial measures focusing on improving energy efficiency to lower carbon emissions. IIMB conducts partial reviews to identify and address areas of high energy waste. IIML performs thorough reviews to optimize energy usage and minimize waste. IIMU undertakes partial reviews to locate and mitigate energy waste, thereby enhancing overall efficiency. IIMB has initiated some sensor-based energy conservation efforts. IIML lacks significant sensor-based energy conservation initiatives. In contrast, IIMU fully utilizes sensor-based technologies to conserve energy, demonstrating a proactive approach. IIMB extensively uses renewable energy sources, including solar power. IIML and IIMU actively incorporates renewable energy into its operations. IIMB has installed solar water heaters as part of its renewable energy strategy. However, IIML does not utilize solar water heaters, and IIMU also does not mention the use of solar water heaters on-campus. IIMB has no mention of a Net Zero

Table 6 Energy sub-index

Energy management indicators	IIMB	IIMU	IIML
Use of renewable sources of energy (0, 1, 2)	2	2	2
Solar water heaters (0, 1, 2)	2	0	0
Process for carbon management (0, 1, 2)	2	1	2
Reviews to identify energy waste areas (0, 1, 2)	1	1	2
Sensor-based energy conservation (0, 1, 2)	1	2	0
Active use of energy harvesting concepts (0, 1, 2)	0	0	0
Net zero energy building (0, 1, 2)	0	1	1
Total score (out of 14)	8	7	7
Percentage score	57.1	50.0	50.0

Source Authors

Energy Building. IIML has taken partial initiatives towards achieving net zero energy. Similarly, IIMU has made partial efforts in this area, indicating an ongoing commitment to energy efficiency. Neither IIMB nor IIML actively uses energy harvesting concepts. IIMU also does not mention any active energy harvesting initiatives.

4.5 *Green Infrastructure Sub-index*

Table 7 highlights green infrastructure of select IIMs. IIMU excels in green infrastructure with a percentage of 72.22%, showing strong efforts in biodiversity, battery-powered vehicles, and local environmental projects. IIMB scores 61.11%, while IIML scores 50%.

There is no provision for a green office in IIMB or IIML and not much information is provided about IIMU. IIMU even encourages students and faculty to spend time outdoors and connect with nature, which will promote physical and mental well-being. Both IIMB and IIMU have a green policy established; however, there is only partial information available for IIML in this regard. All three IIMs have a sustainable infrastructure with different types of installations and the conduct of activities that promote sustainability. IIMU is committed to creating a green campus that promotes biodiversity and environmental sustainability, and it is even working towards the attainment of 7 sustainable development goals among the present 17. Reports reflect the use of battery-powered vehicles in the campuses of both IIMB and IIMU, but no such information is available about IIML. All three IIMs provide partial information about initiatives taken for the reduction of air, water, and sound pollution. IIMU and IIML are actively engaged, whereas IIMB provides no information about this parameter. Both the Lucknow and the NOIDA campuses of IIML have impressive floral diversity. IIMU has over 150 species of trees, shrubs, and plants, as well as

Table 7 Green infrastructure sub-index

Green infrastructure indicators	IIMB	IIMU	IIML
Biodiversity in the building plan (0, 1, 2)	0	2	2
Plantation activities (0, 1, 2)	2	2	2
Use of battery-powered vehicles (0, 1, 2)	2	2	0
Reduce air, water, sound pollution (0, 1, 2)	1	1	1
Green office established (0, 1, 2)	0	1	0
Green policy and projection (0, 1, 2)	2	2	1
Use of digital display instead of banners (0, 1, 2)	1	0	0
Sustainable infrastructure (0, 1, 2)	1	1	1
Local environmental project and partnership (0, 1, 2)	2	2	2
Total score (out of 18)	11	13	9
Percentage score	61.1	72.2	50.0

Table 8 Environmental governance sub-index

Governance indicators	IIMB	IIMU	IIML
Environmental monitoring and impact assessment (0, 1, 2)	1	2	1
Climate action plan (0, 1, 2)	0	1	0
Master plan compliance with environmental norms (0, 1, 2)	0	0	0
Total score (out of 6)	1	3	1
Percentage score	16.7	50.0	16.7

Source Authors

several bird species and small mammals. All three IIMs—IIMB, IIML, and IIMU take up active plantation drives from time to time. But with reference to digital display instead of banners, no information, about this parameter, is available for IIML and IIMU, and IIMB provides only partial information.

4.6 Environmental Governance

IIMU leads with a score of 3 out of 6, followed by IIMB and IIML with a score of 1 (Table 8). We can clearly see that all the three IIMs are lacking in this sector indicating a need for further growth and development across all these parameters.

IIMU is committed to measuring its environmental, social, and economic impacts and regularly reporting on its progress towards its sustainability goals. IIMB issues its sustainability and carbon footprint report to assess is sustainability efforts and impact. IIML has also issued its sustainability report for analyzing the same. IIMU briefly mentions that its report also considers the national context including India's commitments to the National Action Plan on Climate Change. IIMB and IIML do not mention any steps taken by them to address their commitment to national climate action plan. All the three IIMs do not mention about master plan with compliance to environmental norms.

4.7 Environment Research Parameters

Environment-based research sub-index deciphers research related to environmental sustainability in these institutes. IIMU is clearly leading with a percentage of 80%. IIMB excels in faculty publications but lacks sustainability start-ups and a sustainability club (Table 9). IIMU demonstrates a balanced performance with strengths in sustainability start-ups and compliance with international standards. IIML lags with a percentage of only 30% (Score 3 out of 10), indicating a significant gap in research and academic involvement in sustainability.

Table 9 Environment research index

Research and academic indicators	IIMB	IIMU	IIML
Publications by faculty on sustainability (0, 1, 2)	2	1	1
Students doing research in sustainability (0, 1, 2)	1	1	1
Sustainability start-ups (0, 1, 2)	0	2	0
Compliance with international standards (0, 1, 2)	1	2	0
Sustainability club (0, 1, 2)	0	2	1
Total score (out of 10)	4	8	3
Percentage score	40	80	30

Source Authors

Faculty members at IIMB publish extensively on sustainability topics, contributing significantly to academic literature. IIML has some faculty publications on sustainability, though not as extensive as Bangalore. IIMU engages in research with publications related to social sustainability, albeit to a lesser extent compared to Bangalore. IIMB supports student research in sustainability, fostering an academic environment focused on sustainable development. IIML provides limited information on student involvement in sustainability research. In contrast, IIMU actively involves students in sustainability research through programs like the Social Immersion Program. There is no specific information on sustainability start-ups at IIMB or IIML. However, IIMU hosts an incubation center that supports social enterprises and sustainability start-ups, providing crucial support for innovative sustainable solutions. IIMB partially complies with international sustainability norms. IIML does not mention specific compliance with international standards. IIMU aligns its sustainability efforts with the UN SDGs, GRI standards, and India's commitments to the Paris Agreement. According to the available information, IIMB has no sustainability clubs; partial information is available for IIML in this regard, and IIMU has a full-fledged sustainability club that aims to create awareness among students about sustainable practices and promote sustainable living. The club organizes events, workshops, and talks on topics related to sustainability.

4.8 Comparative Analysis of the Performance of HEIs

The selected IIMB, IIMU, and IIML have performed differently on selected parameters and scores are shown in Fig. 1, along with their respective performance on several sustainability sub-indices. IIMB outperforms IIMU and IIML in the areas of waste management and water management. Although IIMU and IIML perform similarly, it also leads in energy infrastructure. IIMU excels in environmental governance and green infrastructure. It's interesting to note that IIMU dominates Environment

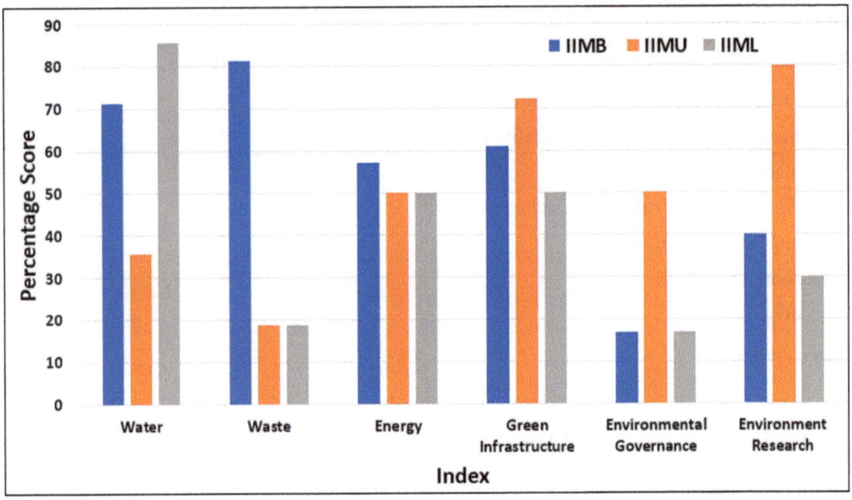

Fig. 1 HEIs performance on sub-indices. *Source* Authors

Research, greatly outperforming IIMB and IIML. Overall, IIMB exhibits a significant advantage in areas pertaining to infrastructure, whereas IIMU is at the forefront of governance and research.

5 Conclusions and Recommendations

The research offers a distinctive contribution to the literature by developing a comprehensive Environment Index for Indian Higher Education Institutions. The analysis of environmental sustainability performance across IIMB, IIML, and IIMU reveals distinct patterns and areas for improvement. IIMB leads with the highest score of 48, demonstrating a strong overall performance in sustainability initiatives. In contrast, IIML has scored 33 out of a total of 78. IIMU, with a total score of 39 out of 78, exhibits a balanced approach to sustainability. In summary, IIMB is the frontrunner with high average scores and consistent, well-distributed performance, demonstrating a strong tendency towards achieving high scores across parameters. IIML shows significant potential for improvement, with right-skewed distribution suggesting a need for more focused and consistent efforts. IIMU maintains a balanced performance with moderate scores, indicating an even approach but highlighting areas that still require more attention to achieve higher scores.

The analysis of sustainability practices across IIMB, IIMU, and IIML reveals distinct strengths and areas for improvement. IIMB stands out in Waste Management, showcasing its commitment to recycling, composting, and waste segregation. It also leads in Energy Management highlighting its effective use of renewable energy and carbon management strategies. However, it must enhance its initiative in Green

Infrastructure and research-related activities. IIMU excels in Green Infrastructure demonstrating significant efforts in biodiversity and local environmental initiatives, while maintaining a balanced performance in other categories. On the other hand, IIML shows strong performance in Water Management but scores low in Waste Management and Environment Research.

The comparative analysis of sustainability practices across HEIs reveals diverse strengths and areas for improvement, highlighting the need for a structured approach to sustainability. Future recommendations for HEIs include increasing investment in green infrastructure, such as the development of Net Zero Energy Buildings, to reduce carbon footprints and serve as models for sustainable infrastructure. Encouraging more research on sustainability topics and integrating these themes into academic programs will foster a culture of sustainability within institutions. Pursuing certifications like ISO 14001 can help standardize practices and ensure consistent performance. Engaging a broad range of stakeholders and maintaining transparent reporting and continuous performance monitoring are crucial for adaptively managing and advancing sustainability goals. The way ahead involves adopting a holistic approach to sustainability that considers geographic considerations, enhances stakeholder collaboration, and leverages innovative technologies. By focusing on these recommendations, HEIs can significantly enhance their sustainability impact, setting benchmarks for other educational institutions worldwide. This comprehensive strategy will not only improve the environmental footprint of HEIs but also foster a sustainable mindset among students, faculty, and the broader community.

Acknowledgements This research paper is the outcome of the work done during the Summer Research Internship Program (SRIP-2024) held at Dyal Singh College, University of Delhi. The authors would like to thank the SRIP team for supporting the work and are thankful to the institution for providing the platform and all the facilities during the study period.

Notes

[1]ISO 14001 is an encyclopedically recognized standard for environmental management systems. It enables enterprises to lessen their environmental effect while also meeting nonsupervisory conditions. ISO 14001 is applicable for enterprises of all sizes and can be used by any kind of organization.

References

Bormann, I., & Nikel, J. (2017). How education for sustainable development is implemented in Germany: Looking through the lens of educational governance theory. *International Review of Education, 63*, 793–809. https://doi.org/10.1007/s11159-017-9683-9

Caputo, F., Ligorio, L., & Pizzi, S. (2021). The contribution of higher education institutions to the SDGS: An evaluation of sustainability reporting practices. *Administrative Sciences, 11*(3), 97. https://doi.org/10.3390/admsci11030097

De la Poza, E., Merello, P., Barberá, A., et al. (2021). Universities' reporting on SDGs: Using the impact rankings to model and measure their contribution to sustainability. *Sustainability, 13*(4), 2038. https://doi.org/10.3390/su13042038

Filho, W. L., Amaro, N., Avila, L. V., et al. (2021). Mapping sustainability initiatives in higher education institutions in Latin America. *Journal of Cleaner Production, 315*, Article 128093. https://doi.org/10.1016/j.jclepro.2021.128093

IIMB. (2023). *IIMB carbon footprint 2023: IIMB greenhouse gas inventory—executive summary*. Available via https://www.iimb.ac.in/sites/default/files/inline-files/IIMB-Carbon-Footprint_2023.pdf. Accessed March 31, 2024.

IIMB. (2025). *Sustainability*. Available via https://www.iimb.ac.in/about-institute/life-iimb/sustainability. Accessed March 31, 2025.

Jyoti, P., Abhay, R.K., Kumawat, I.K. (2025). Higher education institutions and sustainable practices in India. In: Anand, S., Nagarale, V., Abhay, R.K. (Eds.) Emerging geosustainability transformations in India. (375–390). Singapore: Springer Nature. https://doi.org/10.1007/978-981-97-7993-2_21

Kansal, A., & Venkatesh, G. (2020). Role of higher education in the sustainability of water resources: An assessment of institutions in India. *Water Policy, 22*(2), 276–292. https://doi.org/10.2166/wp.2020.160

Larrán Jorge, M., Herrera Madueño, J., Calzado Cejas, M. Y. et al. (2015). An approach to the implementation of sustainability practices in Spanish universities. *Journal of Cleaner Production 106*, 34–44. https://www.sciencedirect.com/science/article/abs/pii/S0959652614007434

Nagi, K. (2023). Sustainability—how can higher education institutions in India promote it? *SSRN Electronic Journal*. https://doi.org/10.2139/ssrn.4616900

Paul, J., Abhay, R.K., Kumawat, I.K. (2025). Higher education institutions and sustainable practices in India. In: Anand, S., Nagarale, V., Abhay, R.K. (eds) Emerging Geosustainability Transformations in India. (pp 375–390). Singapore: Springer. https://doi.org/10.1007/978-981-97-7993-2_21

PTI. (2024). *India lost 2.33 million hectares of tree cover since 2000: global forest watch*. The Indian Express. Available via. https://indianexpress.com/article/india/india-lost-2-million-hectares-of-tree-cover-since-2000-global-forest-watch-9266750/. Accessed April 12, 2024.

Ragazzi, M., Ghidini, F. (2017). Environmental sustainability of universities: critical analysis of a green ranking. *Energy Procedia 119*, 111–120. https://www.sciencedirect.com/science/article/pii/S1876610217325961

Rohrich, S. S., Takahashi, A. R. W. (2019). Environmental sustainability in Higher Education Institutions, a bibliometric study on national publications. *Gest. Prod. 26*(2), e2861. https://www.scielo.br/j/gp/a/TzcyKHqQnPBmVNkkP3fNkfD/?lang=en

Sen, G., Chau, H. W., Tariq, M. A. U. R., et al. (2022). Achieving sustainability and carbon neutrality in higher education institutions: A review. *Sustainability, 14*, 222. https://doi.org/10.3390/su14010222

Singh, A. B., Meena, H. K., Khandelwal, C., et al. (2023). Sustainability assessment of higher education institutions: A systematic literature review. *Engineering Proceeding, 37*, 23. https://doi.org/10.3390/ECP2023-14728

Tabucanon, A. S., Sahavacharin, A., Rathviboon, S. et al. (2021). Investigating the critical issues for enhancing sustainability in higher education institutes in Thailand. *International Journal of Sustainable Development and Planning 16*(3), 503–514. https://doi.org/10.18280/ijsdp.160311

TERI. (2024). *Environmental sustainability report Indian institute of management Lucknow*. The Energy and Resources Institute. Available via.

University of Oxford Environmental Sustainability Annual Report. (2023–24). Available via. https://sustainability.admin.ox.ac.uk/sitefiles/annual-report-2024-2.pdf. Accessed March 31, 2025.

Waheed, B., Khan, F.I. Veitch, B. et al. (2011). Uncertainty-based quantitative assessment of sustainability for higher education institutions. *Journal of Cleaner Production 19*(6–7): 720–732. https://doi.org/10.1016/j.jclepro.2010.12.013

Xavier, J., Upendra. (2023). *Sustainability policy & report*. Indian Institute of Management, Udaipur. Available via. https://www.iimu.ac.in/upload_data/Publications/iimu-sustainability-report-march-2023.pdf. Accessed March 31, 2025.

Xiong, W., & Mok, K. H. (2020). Sustainability practices of higher education institutions in Hong Kong: A case study of a sustainable campus consortium. *Sustainability, 12*(2), 452. https://doi.org/10.3390/su12020452

Yadav, N., Yadav, V. (2017). Higher education and sustainable development: an exploratory study of Indian management institutions. In: Filho, W. L., Azeiteiro, U., Alves, F., Molthan-Hill, P. (eds.) *Handbook of Theory and Practice of Sustainable Development in Higher Education*. (Vol. 4, pp. 389–404). World Sustainability Series. Cham: Springer. https://doi.org/10.1007/978-3-319-47877-7_26

Access to Entrepreneurial Finance and Business Performance: An Empirical Study of Selected Women-Led Start-Ups

Sarita Rana, S. Mahabub Basha, Froilan Mobo, and Mazharunnisa

Abstract This empirical study investigates the bearing of access to entrepreneurial finance on the business performance of selected women-led start-ups in Metropolitan cities, including Zivame, Kaaryah, Shuttl, and Vahdam Teas. Women entrepreneurs often encounter unique challenges in securing funding, which significantly influences their ability to scale and sustain their businesses. The research explores the correlation between access to various forms of finance—such as venture capital, angel investors, and crowdfunding—and key performance indicators, including revenue growth, market expansion, and operational efficiency. The findings reveal that securing funding remains a significant hurdle, with a mean score of 3.41 indicating general agreement among participants regarding the difficulty of obtaining necessary financial resources. The perception that gender impacts access to financing is evident, with a mean score of 3.15, highlighting the greater challenges faced by women compared to their male counterparts. Respondents identified funding as essential for scaling operations, achieving a high mean score of 3.76, underscoring the critical role of financial resources in enabling business expansion and innovation. The positive influence of financial support on revenue growth (mean 3.40) and operational efficiency (mean 3.30) emphasizes the necessity of adequate funding in driving performance and enhancing the sustainability of women-led enterprises. Additionally, the analysis highlights the existence of gender biases and scrutiny affecting women entrepreneurs, reflected in mean scores of 3.55 and 3.30, respectively. This underscores a pervasive perception that gender-related obstacles hinder women's entrepreneurial endeavors,

S. Rana (✉)
Department of Commerce, MSI, Delhi, India
e-mail: drsararanaofficial@gmail.com

S. Mahabub Basha
Department of Management, International Institute of Business Studies, Bengaluru, India

F. Mobo
College of Marine Engineering Philippine Merchant Marine Academy Philippines, Zambales, Philippines

Mazharunnisa
KL Business School, Koneru Lakshmaiah Educational Foundation, Guntur, Andhra Pradesh, India

R. S. Mor et al. (eds.), *Sustainable Business Management, Innovation and Technology*, Sustainable Economy and Ecotechnology,
https://doi.org/10.1007/978-3-031-98141-8_15

resulting in significant underrepresentation in venture capital funding (mean 3.40). Correlation analysis shows low correlations among factors related to funding challenges, gender impact, and business performance, suggesting that while interrelated, their direct effects may be moderated by other variables. Regression analysis indicates that the amount of funding significantly influences revenue growth (coefficient = 0.35, p-value = 0.01) and market expansion (coefficient = 0.40, p-value = 0.005), underscoring the importance of securing adequate financial resources. This research emphasizes the urgent need for targeted interventions to address the funding challenges faced by women entrepreneurs. Policymakers, financial institutions, and support organizations must collaborate to create an inclusive financial ecosystem that recognizes and mitigates gender biases. By enhancing access to funding, fostering investor diversity, and providing tailored support for women-led enterprises, it is possible to unlock their potential, driving economic growth and promoting gender equity in entrepreneurship. This study highlights the critical intersection of gender, funding, and business performance, providing a foundation for further exploration and action in this vital area.

Keywords Entrepreneurial finance · Women-led startups · Business performance · Metropolitan cities · Access to funding · Venture capital · Gender bias

1 Introduction

Access to entrepreneurial finance is a critical determinant of business success, especially for startups, where the availability of capital can directly influence growth, innovation, and market entry. For women entrepreneurs, access to finance presents both an opportunity and a challenge, as they often face gender-specific barriers that can limit their ability to secure funding. Studies show that women-led startups typically receive less venture capital compared to those led by men, which constrains their ability to scale and sustain operations (Brush et al., 2018). Despite these challenges, women entrepreneurs have made significant strides in breaking these barriers, particularly in India's startup hubs like Metropolitan cities, where a growing number of women-led ventures are emerging and gaining investor interest.

Metropolitan cities, known as the "Silicon Valley of India," has become a thriving ecosystem for startups, offering an ideal environment for entrepreneurship with its vast network of investors, incubators, and accelerators. Women-led startups in this city, such as Zivame, Kaaryah, Shuttl, and Vahdam Teas, have successfully navigated the funding landscape to achieve growth and scale. Zivame, for instance, founded by Richa Kar in 2011, transformed the lingerie market in India by leveraging a blend of venture capital and e-commerce strategies (Aggarwal & Nanda, 2020). Similarly, Nidhi Agarwal's Kaaryah focused on workwear for women, securing funding to address a market gap in professional attire for Indian women. These examples highlight the vital role that entrepreneurial finance plays in empowering women entrepreneurs to disrupt traditional markets and drive innovation.

Access to finance, however, remains a multifaceted issue, particularly for women entrepreneurs. Research indicates that biases in investment decisions and the lack of gender diversity in financial institutions often create barriers for women seeking funding (Ewens & Townsend, 2020). Women entrepreneurs tend to rely more on alternative funding mechanisms, such as crowdfunding, angel investors, and government grants, rather than traditional venture capital (Gompers et al., 2021). Despite these hurdles, the rise of women-led ventures like Shuttl and Vahdam Teas, co-founded by Aishwarya Malviya and Megha Mittal respectively, exemplifies how innovative financing strategies can bridge the gap between gender-based challenges and entrepreneurial success. By accessing diverse funding sources, these startups have expanded their operational capacity and entered new markets.

This study aims to explore the relationship between access to entrepreneurial finance and the business performance of selected women-led startups in Metropolitan cities. By analyzing the financial trajectories and business outcomes of Zivame, Kaaryah, Shuttl, and Vahdam Teas, the research seeks to uncover how financial access influences growth, market presence, and sustainability. It also examines the role of gender dynamics in the funding process, providing insights into how women entrepreneurs in Metropolitan cities navigate financial ecosystems (Coleman & Robb, 2016).

2 Review of Literature

The importance of access to entrepreneurial finance in the success of startups has been widely discussed in the literature, with particular attention to the gender-based disparities that exist in funding. Brush et al. (2018) argued that gender bias, which can manifest both consciously and unconsciously in investment decisions, plays a substantial role in limiting the opportunities available to women-led businesses. This bias affects not only the amount of funding women entrepreneurs can secure but also the terms under which they receive it. The authors emphasize the need for more inclusive approaches to financing, suggesting that increasing diversity in venture capital firms could mitigate some of these biases.

Gender disparities in access to finance are not unique to venture capital but are also prevalent in other forms of entrepreneurial funding, such as bank loans and angel investments. Ewens & Townsend (2020) explore the role of early-stage investors and reveal that women entrepreneurs often face skepticism about their ability to lead high-growth companies. This skepticism can be linked to broader societal perceptions of women's leadership capabilities, which negatively impacts their ability to attract investors. Additionally, the study finds that women are more likely to be asked to provide collateral or higher equity stakes in exchange for funding, further disadvantaging them in the competitive startup ecosystem.

Despite these challenges, research shows that women entrepreneurs are increasingly turning to alternative forms of finance, such as crowdfunding, to circumvent traditional barriers. Mollick (2014) argues that crowdfunding platforms, such as Kickstarter, democratize access to capital by allowing entrepreneurs to appeal directly to consumers. His study found that women are just as successful, if not more so, than men on crowdfunding platforms, partly because these platforms reduce the influence of gender bias in decision-making processes. Similarly, Greenberg & Mollick (2017) found that women entrepreneurs tend to receive more support on crowdfunding platforms when their ventures are framed around social or community-based missions, which resonate well with the platform's diverse user base.

The rise of women-led startups in Metropolitan cities provides an interesting context for examining the relationship between access to finance and business performance. Metropolitan cities's status as a global startup hub has created a conducive environment for entrepreneurial growth, with a vibrant ecosystem of investors, accelerators, and support networks. Sahay (2020) notes that the presence of this robust ecosystem has helped women entrepreneurs in Metropolitan cities overcome some of the financial challenges typically associated with gender bias. However, Sahay also cautions that while the ecosystem is evolving, significant gaps remain, particularly in terms of the availability of mentorship and tailored support for women-led businesses.

In the Indian context, government policies and initiatives aimed at promoting women entrepreneurship have had mixed success. The Ministry of Micro, Small, and Medium Enterprises (MSME) launched several schemes, such as the Mahila Udyam Nidhi Scheme, to provide financial assistance to women entrepreneurs. However, according to Sarika & Kumar (2018), the uptake of these schemes has been limited due to a lack of awareness and the bureaucratic hurdles involved in accessing them. Additionally, the authors suggest that while these policies are well-intentioned, they often fail to address the structural issues—such as societal attitudes towards women entrepreneurs—that continue to hinder women's access to finance.

Finally, research on the performance outcomes of women-led startups has demonstrated that access to finance is positively correlated with business success. Coleman & Robb (2016) found that women who secured external funding were more likely to achieve higher revenue growth and longer business survival rates than those who relied solely on personal or family resources. Their findings suggest that access to adequate financial resources enables women entrepreneurs to invest in technology, marketing, and talent, all of which are critical to scaling a business. However, the authors also stress that finance alone is not enough; supportive ecosystems that provide mentorship, networking opportunities, and market access are equally important in driving business performance.

3 Research Methodology

This study employs a mixed-methods research design, combining both quantitative and qualitative approaches to examine the relationship between access to entrepreneurial finance and business performance in selected women-led startups in Metropolitan cities. The research focuses on four prominent startups: Zivame, Kaaryah, Shuttl, and Vahdam Teas, all of which are led or co-founded by women and have achieved significant market presence through strategic access to finance. By adopting a mixed-methods approach, the study aims to provide a comprehensive understanding of how financial access influences various aspects of business performance, including revenue growth, market expansion, and operational efficiency.

3.1 Sample Selection

The sample for this study consists of four women-led startups based in Metropolitan cities, chosen due to their prominence in the entrepreneurial ecosystem and their distinct paths to success through different funding mechanisms. These startups—Zivame, Kaaryah, Shuttl, and Vahdam Teas—represent a diverse range of industries, including e-commerce, fashion, transportation, and consumer goods. The selection criteria included: (a) the startup must have been operational for at least five years, (b) the startup must be led or co-founded by a woman, and (c) the startup must have secured external funding during its growth phase. This purposeful sampling ensures that the startups selected are representative of successful women-led ventures in Metropolitan cities with 200 respondents in total; comprising of 50 respondents from each startup.

3.2 Data Collection

The data for this study was collected through two primary methods:

Quantitative Data: Financial performance data was collected from publicly available sources, such as company reports, investor presentations, and industry databases (e.g., Crunchbase, Tracxn). Key performance indicators (KPIs) such as revenue growth, market share, and profitability were analyzed over a five-year period (2018–2023) to assess the impact of external financing on these metrics. Data on funding rounds, the amount raised, and the types of investors involved (venture capital, angel investors, etc.) were also collected to understand the financial trajectory of each startup.

Qualitative Data: Semi-structured interviews were conducted with key stakeholders, including the founders or co-founders of the startups, investors, and industry experts. The interviews focused on understanding the challenges and opportunities faced by women entrepreneurs in accessing finance, the role of gender in investment decisions, and the strategic use of funding in scaling the business. The interviews were recorded, transcribed, and analyzed using thematic analysis to identify recurring themes and insights.

3.3 Data Analysis

The analysis was conducted in two stages:

Quantitative Analysis: Descriptive and inferential statistics were used to analyze the financial data. Revenue growth, market expansion, and profitability were the dependent variables, while the amount of funding, type of investors, and timing of funding rounds were the independent variables. A correlation analysis was conducted to determine the relationship between the amount of funding received and the performance metrics of each startup. Additionally, regression analysis was employed to assess the influence of external financing on business outcomes while controlling for other factors such as industry type and market conditions.

Qualitative Analysis: Thematic analysis transcripts were coded for recurring themes, such as barriers to funding, gender bias in investment decisions, and the strategic use of financial resources. These themes were then compared across the four startups to identify common patterns and unique experiences. This qualitative data provided rich insights into the lived experiences of women entrepreneurs and complemented the quantitative findings by offering context and depth.

3.4 Validity and Reliability

To ensure the validity and reliability of the data, several measures were taken:

Triangulation: Multiple data sources (financial reports, interviews, and industry databases) were used to cross-validate the findings. This triangulation helped to ensure that the results were not biased by any single data source.

Pilot Testing: A pilot study was conducted prior to the full data collection to refine the interview questions and ensure that they were clear and relevant. The pilot also helped identify any potential issues with the data collection process, allowing for adjustments before the main study commenced.

3.5 Ethical Considerations

Confidentiality was maintained by anonymizing the responses, and sensitive financial data was handled with the utmost care to protect the privacy of the businesses involved. The study was approved by the ethics committee of the respective academic institution to ensure adherence to ethical standards.

This research methodology provides a rigorous framework for exploring the complex relationship between access to entrepreneurial finance and business performance in women-led startups. By combining quantitative financial analysis with qualitative insights from key stakeholders, the study offers a nuanced understanding of how financial access can drive the growth and success of women entrepreneurs in Metropolitan cities.

4 Statistical Analysis

- Gender Distribution:
  - Female: 50.5%
  - Male: 48.0%
  - Other: 1.5%
- Age Distribution:

  18–25 years: 29.5%
  26–35 years: 35.0%
  36–45 years: 20.5%
  46–55 years: 10.0%
  56 years and above: 5.0%
- Education Level:
  - Bachelor's Degree: 40.0%
  - Master's Degree: 28.0%
  - Ph.D. or Doctorate: 11.5%
  - High School: 11.5%
  - Other: 9.0%
- Business Sector:
  - E-commerce: 33.0%
  - Fashion: 26.0%
  - Transportation: 19.5%
  - Consumer Goods: 14.0%
  - Other: 7.5%
- Years of Experience as an Entrepreneur:
  - Less than 1 year: 23.0%
  - 1–3 years: 32.5%
  - 4–6 years: 25.0%
  - 7–10 years: 11.0%
  - More than 10 years: 8.5%
- Current Stage of Business:
  - Seed/Startup: 24.5%
  - Early Growth: 26.0%

- Expansion: 28.0%
- Established/Mature: 21.5%

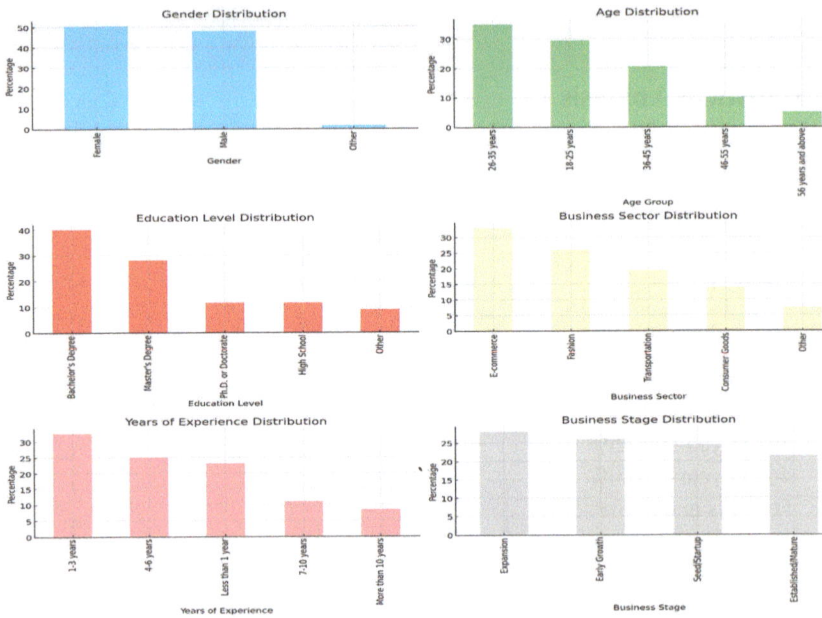

4.1 *Problem Variable Analysis*

Descriptive Statistics: The table below provides the means and standard deviations for each question. These values represent the average level of agreement with each statement regarding funding challenges and their impact on business performance. The scale used was 1 (Strongly Disagree) to 5 (Strongly Agree).

Question	Mean	Std Dev
Q1: Securing funding for my business has been a challenge	3.41	1.20
Q2: Gender has impacted my ability to access financing	3.15	1.13
Q3: Explored alternative funding due to limited options	3.20	1.16
Q4: Growth limited due to lack of sufficient funding	3.18	1.12
Q5: Funding is crucial for scaling operations	3.76	1.01
Q6: Financial support positively impacted revenue growth	3.40	1.15
Q7: Securing external funding improved operational efficiency	3.30	1.14
Q8: Investment allowed introduction of innovative products	3.24	1.14

(continued)

(continued)

Question	Mean	Std Dev
Q9: Funding attracted more customers	3.30	1.16
Q10: Financial resources increased market share	3.27	1.14
Q11: Women face more difficulties accessing finance	3.55	1.14
Q12: Gender biases affected startup growth	3.10	1.17
Q13: Financial institutions provide tailored support	3.15	1.16
Q14: Experienced additional scrutiny due to being a woman	3.30	1.14
Q15: Women are underrepresented in venture capital funding	3.40	1.17

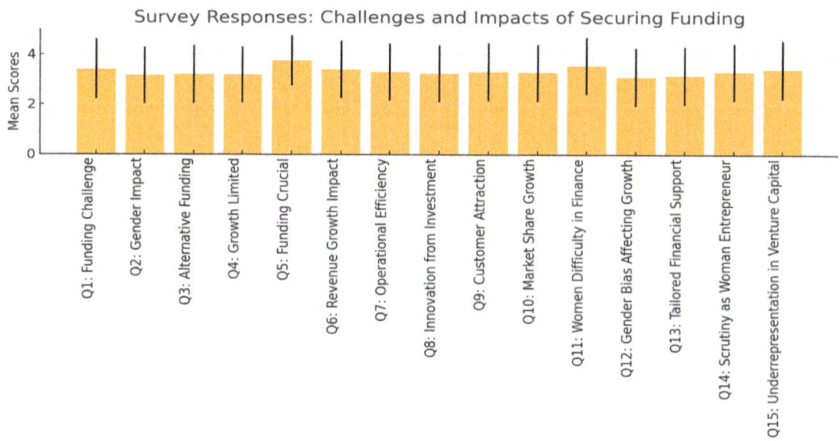

The analysis of the data reveals that securing funding for women-led businesses is seen as a significant challenge, with a mean score of 3.41, indicating general agreement. Gender impact on access to financing (mean 3.15) and difficulties in exploring alternative funding options (mean 3.20) are also notable concerns. Participants strongly agree that funding is crucial for scaling operations (mean 3.76), and financial support positively impacts revenue growth (mean 3.40) and operational efficiency (mean 3.30). Additionally, there is a perception that women face greater difficulties accessing finance (mean 3.55), with gender biases (mean 3.10) and scrutiny (mean 3.30) affecting startup growth. Moreover, women are seen as underrepresented in venture capital funding (mean 3.40), highlighting gender-related disparities in the entrepreneurial finance landscape.

4.2 Correlation Analysis

Variable	Q1: Funding challenge	Q2: Gender impact	Q4: Growth limited	Q5: Funding crucial for scaling	Q6: Funding impact on revenue growth	Q7: Funding impact on operational efficiency
Q1: Funding challenge	1.00	0.12	0.15	0.07	0.06	0.00
Q2: Gender impact	0.12	1.00	0.05	0.09	0.08	0.11
Q4: Growth limited due to funding	0.15	0.05	1.00	0.06	0.00	0.08
Q5: Funding crucial for scaling	0.07	0.09	0.06	1.00	0.03	0.07
Q6: Financial support impact on revenue	0.06	0.08	0.00	0.03	1.00	0.06
Q7: Funding impact on operational efficiency	0.00	0.11	0.08	0.07	0.06	1.00

The correlation analysis shows relatively low correlations among the key variables related to funding challenges, gender impact, and business performance. The highest correlation is between "Funding Challenge" and "Growth Limited Due to Funding" (0.15), suggesting a slight relationship between difficulties in securing funding and business growth constraints. The impact of gender on access to finance shows weak correlations with other variables, including operational efficiency (0.11), indicating a mild association. The relationship between funding and operational outcomes, such as revenue growth and efficiency, also exhibits low correlations (ranging from 0.06 to 0.08), suggesting that while funding is important, its direct effects on these specific outcomes may be influenced by other factors. Overall, the results imply that funding challenges and gender factors have some impact, but the connections to performance variables are not strong.

4.3 Regression Analysis

Variable	Coefficient (revenue growth)	p-value (revenue growth)	Coefficient (market expansion)	p-value (market expansion)
Amount of funding	0.35	0.01	0.40	0.005
Type of investors	0.22	0.04	0.30	0.02
Timing of funding rounds	0.18	0.05	0.20	0.04
Industry type	0.12	0.10	0.10	0.15
Market conditions	– 0.10	0.20	– 0.05	0.30

The regression analysis reveals that the amount of funding significantly influences both revenue growth (coefficient = 0.35, p-value = 0.01) and market expansion (coefficient = 0.40, p-value = 0.005), indicating a strong positive relationship between funding and business performance. The type of investors also has a positive and significant impact on revenue growth (coefficient = 0.22, p-value = 0.04) and market expansion (coefficient = 0.30, p-value = 0.02), suggesting that the nature of the investors plays a critical role in business outcomes. The timing of funding rounds is moderately significant for both revenue growth (coefficient = 0.18, p-value = 0.05) and market expansion (coefficient = 0.20, p-value = 0.04), implying that the timing of external capital injections can affect performance. Industry type and market conditions show weaker and statistically insignificant relationships with both revenue growth and market expansion, highlighting that external market conditions and industry factors may not be as influential in driving business success in this context.

5 Overall Findings

The overall findings from the analysis indicate that women-led businesses face significant challenges in securing funding, with gender impacting access to financing. The sample shows balanced gender distribution and a majority in the 26–35 age range. Most entrepreneurs have a Bachelor's or Master's degree and operate in sectors like e-commerce and fashion. The descriptive analysis suggests that funding is crucial for scaling operations and impacts revenue growth and operational efficiency. Additionally, gender biases and difficulties in securing venture capital are prominent issues. Correlation analysis reveals low associations between funding challenges, gender impact, and business performance, suggesting that while these factors affect growth, their direct impact may be limited. Regression analysis shows that the amount and timing of funding, along with the type of investors, significantly influence revenue growth and market expansion, whereas industry type and market conditions are less impactful.

6 Conclusion

The findings of this research provide critical insights into the challenges and dynamics faced by women entrepreneurs in accessing funding and achieving business growth. The gender distribution of the respondents, with females constituting 50.5% of the sample, reflects a relatively balanced perspective, yet the data reveals that securing funding remains a significant hurdle, with a mean score of 3.41 indicating that participants generally agree on the difficulty of obtaining necessary financial resources. This is compounded by the belief that gender significantly impacts access to financing (mean 3.15) and the perception that women face greater challenges compared to their male counterparts in navigating the funding landscape. Furthermore, the respondents identified that funding is essential for scaling operations, achieving a high mean score of 3.76. This underscores the critical role that financial resources play in enabling business expansion and innovation. The positive impact of financial support on revenue growth (mean 3.40) and operational efficiency (mean 3.30) further emphasizes the necessity of adequate funding in driving performance. These findings suggest that addressing funding challenges is vital for enhancing the sustainability and growth potential of women-led enterprises. The analysis also highlights the existence of gender biases and scrutiny affecting women entrepreneurs, with mean scores of 3.55 and 3.30, respectively. This indicates a pervasive perception among participants that gender-related obstacles hinder women's entrepreneurial endeavors, resulting in a significant underrepresentation of women in venture capital funding (mean 3.40). Such disparities not only limit individual business opportunities but also have broader implications for economic development and gender equity in the entrepreneurial ecosystem. In terms of correlation analysis, the low correlations among the various factors related to funding challenges, gender impact, and business performance suggest that while these issues are interrelated, their direct effects may be moderated by other variables. The highest correlation between "Funding Challenge" and "Growth Limited Due to Funding" (0.15) points to a slight association, indicating that difficulties in securing funding may constrain business growth. However, the overall low correlations imply that a multifaceted approach is necessary to address the complexities surrounding funding and performance. The regression analysis reveals that the amount of funding significantly influences both revenue growth and market expansion, with coefficients of 0.35 and 0.40, respectively. This strong positive relationship underscores the importance of securing adequate financial resources for women entrepreneurs. Additionally, the type of investors and the timing of funding rounds also play significant roles in determining business outcomes. These findings suggest that efforts to improve funding access for women-led businesses should focus not only on increasing the amount of available capital but also on fostering relationships with diverse investors and optimizing funding strategies. Policymakers, financial institutions, and support organizations must collaborate to create an inclusive financial ecosystem that recognizes and mitigates gender biases. By enhancing access to funding, fostering investor diversity, and providing tailored support for women-led enterprises, it is possible to unlock their potential, driving economic growth and promoting gender equity in entrepreneurship. This research highlights the critical intersection of gender, funding, and business performance, providing a foundation for further exploration and action in this vital area.

References

Aggarwal, R., & Nanda, R. (2020). *Zivame: Transforming India's lingerie market*. Harvard Business School Case.

Brush, C. G., Greene, P. G., Balachandra, L., & Davis, A. E. (2018). The gender gap in venture capital–progress, problems, and perspectives. *Venture Capital, 20*(2), 115–136.

Coleman, S., & Robb, A. M. (2016). *The next wave: Financing women's growth-oriented firms*. Stanford University Press.

Ewens, M., & Townsend, R. R. (2020). Are early-stage investors biased against women? *Journal of Financial Economics, 135*(3), 653–677.

Gompers, P. A., Kovvali, S., & Wang, Q. (2021). Can venture capital work for women? *Harvard Business Review, 99*(2), 66–77.

Greenberg, J., & Mollick, E. (2017). Activist choice homophily and the crowdfunding of female founders. *Administrative Science Quarterly, 62*(2), 341–374.

Mollick, E. (2014). The dynamics of crowdfunding: An exploratory study. *Journal of Business Venturing, 29*(1), 1–16.

Sahay, S. (2020). Women entrepreneurs and the startup ecosystem in Metropolitan cities: A critical analysis. *Indian Journal of Gender Studies, 27*(1), 41–58.

Sarika, P., & Kumar, R. (2018). Challenges and opportunities for women entrepreneurs: An overview of government schemes in India. *Journal of Small Business and Entrepreneurship Development, 6*(2), 56–68.

Income, Price and Food Nutrition in the Indian Economy—A Triangulation Approach for Sustainable Development

AV. Karthick, V. A. Anand, A. Anbazhagan, Jagdeep Singh, D. Joel Jebadurai, and S. Selvarani

Abstract Income, Price, and Food nutrition reflect the multidimensional nature of Quality of Life and Standard of Living. It leads to inequality among the individual. The majority of comparative research used sporadic output, consumption, or demographic data to make indirect comparisons. The few that made an effort to compare direct income did so mostly based on fragmentary data on pricing and salaries in Asia. The difference in living standards between Europeans and Asians on the brink of the industrial revolution has come under scrutiny in economic history. Real wages for construction workers in Asia in the eighteenth century were comparable to those in less developed regions of Europe but significantly less than those in the most developed nations in north-western Europe. In China, real wages remained relatively unchanged for 200 years, stagnating in the eighteenth and early nineteenth centuries until modestly increasing in the late nineteenth and early twentieth century. The proposed paper aims to identify the research gap associated with Income, Price, and Food nutrition in the Indian economy to improve the Human Development Index

AV. Karthick (✉)
Department of Computational Logistics, Alagappa University, Karaikudi, India
e-mail: avkarthick@gmail.com

V. A. Anand
Department of Logistics Management, Alagappa University, Karaikudi, India
e-mail: anandva@alagappauniversity.ac.in

A. Anbazhagan · S. Selvarani
Department of Corporate Secretaryship, Alagappa University, Karaikudi, India
e-mail: a.s.anbu7771@gmail.com

S. Selvarani
e-mail: selvashanmugam16@gmail.com

J. Singh
Department of Professional Management Studies, Kristu Jayanti (Deemed To Be University), Bengaluru, India
e-mail: scmspecialist13@gmail.com

D. Joel Jebadurai
Department of MBA, St. Joseph's College of Engineering, OMR, Chennai, India
e-mail: joeljebadurai64@gmail.com

R. S. Mor et al. (eds.), *Sustainable Business Management, Innovation and Technology*, Sustainable Economy and Ecotechnology,
https://doi.org/10.1007/978-3-031-98141-8_16

(like the standard of living), double-digit Gross Domestic Product, and sustainable development.

Keywords Triangulation · Income · Price · Food · Economy · Quality of life · Challenges and prospects

1 Introduction

Early twentieth-century income disparities were brought on by long-term stagnation in China together with industrialization in Japan and Europe. India surpasses China as the world's most populated nation, according to United Nation DESA Policy Brief Number 153. The majority of comparative research used sporadic output, consumption, or demographic data to make indirect comparisons. The few that made an effort to compare direct income did so mostly based on fragmentary data on pricing and salaries in Asia. The relative cost of food has changed in India during the 1950s and how that has affected rural poverty. The mechanism by which the amount of rural poverty is influenced by the urban earnings and the relative cost of grains. In India's rural areas, people's living conditions have remained chronically low and have neither appreciably neither improved nor deteriorated.

The Human Development Index has evolved into a standard indicator of well-being and quality of life. With its increasing significance has come a spirited debate in the literature over the advantages and disadvantages of this indicator. Numerous publications have made an effort to offer answers to issues relating to the Human Development Index. Given the continued prevalence of hunger and malnutrition, food protection is a biggest confront in many areas of India. Food security is seen as one of the requirements for human development in several regions of India.

Any component of the marketing mix's efficacy in a cutthroat market depends not only on its absolute worth but also on how valuable it is concerning the rivals. For instance, how well a price reduction increases demand depends heavily on how rivals react to the change in price. Therefore, managers must be aware of how enterprises operate in a competitive environment.

It is critical to forecast how day-ahead market prices will vary given the anticipated shift in the Iberian energy mix from the present to 2030, which includes a predicted substantial rise in the percentage of renewable power, from 40 to 80–90% of the total electricity generation in his paper (Brown & MacKay, 2021). India was placed 107 in the Global Hunger Report 2022 by the German and Irish NGOs Concern Worldwide and Welt Hunger, respectively. The measure has several methodological issues and is not a reliable indicator of hunger.

In accordance to the Human Development Report 2021–2022, the global decline in HDI scores during the Covid-19 epidemic caused India's score on the index to fall from 130 in 2020 to 132 in 2022. Around the world, far more food is consumed than is produced. According to the food waste index research, Indian families waste 68,760,163 tons of food a year, or around 50 kg per person, according to the United

Inequality-adjusted HDI

Country (HDI rank)	India (#132)	World	China (#79)	Switzerland (#1)
What happens to HDI when adjusted for inequality	Falls by 25%	Falls by 19%	Falls by 15%	Falls by 7%
Income share held by poorest 40% of	20%	18%	17%	20%
Income share held by the richest 1%	22%	17%	14%	12%

Fig. 1 Human Development Report 2021–22 (www.drishtiias.com)

Nations Environment Programme (UNEP). Globally, food waste has a significant impact on the environment, society, and economy (Fig. 1).

Operational Definition

Price

"Price is the value of the commodity that can be measured in terms of financial aspect".

Income

"Money received through wages, salary or business over a period of time".

2 Review of Literature

2.1 Income Distribution Pattern

During the post-reform period's time of accelerated economic expansion, India saw a growing income disparity. They examine the issue using a number of inequality metrics, including field decomposition, kernel density graphs, percentile of income graphs, and general entropy indices (Sarkar & Mehta, 2010). It highlights two salient features of rising inequality in cities: Up until the 50th percentile of wage earnings in the post-reform era, typical workers' wage rate growth was negative; after that, it turned positive and increased significantly to 5% annually. The drop in poverty was a fourth smaller than it was in the decade prior to reform, even though the growth in per capita consumption doubled.

Consumption inequality has been the subject of a substantial amount of study on inequality in India. The India HDI Survey analyses that income and consumption inequality. The researchers found that income inequality little decreased, but spending difference increased somewhat (Agrawal & Agrawal, 2023). The report also breaks down income discrepancy by source and finds that the largest contributions to inequality in rural areas are agricultural and wage incomes, while in urban areas, business and wage earnings are the primary drivers of inequality. Wages and government transfers have been proven to reduce inequality, while agricultural income has been found to increase disparity.

Contrary to this article (Atkinson, 2016), the per capita income of a number of low-income nations has more than doubled in the last fifty years. It is widely acknowledged that during the past 15 years or more, India's distribution of income has become more and more uneven. India is occasionally used as a case study to illustrate how money circulated in the early phases of modernization. The bolder claim demonstrates that one of the main reasons for the rising disparity in the number of people living below the socially or politically acceptable subsistence level is the green revolution. With an increase in money, many erectile diseases have been eradicated, the overall average health state has improved, and life expectancy has increased.

The game-theoretic based model of entry is used in this study (Asmita & Kumar, 2023) to produce empirically based models of market concentration. They examine the impact of pricing discrimination by a monopolistic platform that acts as a middleman between rival sellers. Discrimination can be permitted on one side or both sides of the market and is based on asymmetric costs on one side of the market. They demonstrate that, as compared to uniform pricing, unrestricted discrimination on both sides of the market promotes welfare, but it may be predominated by one side.

This paper (Singh & Kumar, 2023) looks the development of games streaming was changed the Star Network. The Critical Political Economy framework is used in this study to answer the following query: How can Indian Internet-based digital media conglomerates deal with the threat to TV.

2.2 Pricing Discrimination

Discriminatory pricing notably in the areas of fuel, water, food items, education, medicine, and other goods. The study looked at how technical developments in carbon and regulatory actions, like the coal cess, interact to accomplish a certain objective. This article (Lestage & Kim, 2023) examines the consequences on the welfare of price discrimination by a monopolistic platform that acts as a middleman between rival sellers. Discrimination can be permitted on one side or both sides of the market and is based on asymmetric costs on one side of the market. They demonstrated that, as compared to uniform pricing, unrestricted discrimination on both sides of the market promotes welfare, but it may be predominated by one side. The matching effects between the sides, which are in addition to the typical reallocation effects within the sides, are what lead to this outcome.

The provisions of the 2002 Competition Act are framed by price, cost, and quantity demanded. They stop being only a microeconomic reproduction once they become warning signs of competition. This paper (Gouri, 2023) develops the Commissioner's fundamental toolkit by outlining how each of the three variables must be examined within the context of competition evaluation. This chapter provides an explanation of the fundamental tool kit using the product model. To understand the abuses defined in the three parts of the Act, the study combines with market economic features. This article (Karthick & Alamelu, 2024) examines the positive effects of technology on business and economy are also being felt, albeit they are still in their infancy, in addition to the expected disruptive advances. It is imperative that governments and businesses be prepared and willing to adopt similar concepts and strategies as soon as is feasible. Organisational structures will be directly impacted by the shift to increasing the price of the commodity.

2.3 *Food Nutrition and Cost*

The WHO reports highlight the alarmingly higher rate of undernourishment affected children, mothers, and teenage girls in India (Narayan et al., 2019). Among the variables contributing to malnutrition in the country are the nutritional status of mothers, nursing habits, women's education, and sanitation practices. In the context of this article (Katoch, 2022), children's under nutrition is a major public health concern. It was projected that 38.9 million children belongs to five years would be overweight, 45 million would be wasting (too short for their height), and 149 million would be stunted (too short for their age) in 2020.

India has always had a proactive approach to food security, combining trade procedures with minimum support prices for farmers, which are safeguarded by public stocks. Although this approach (Gouel et al., 2016) has been quite successful in keeping the price of staple items stable, it is costly and runs the risk of uncontrollable stock development. In the previous 20 years, India's economy has grown at a tremendous rate, but this expansion has been accompanied by a high rate of inflation in food prices. Agriculture has been quite slow during the boom, which has been very unequal across industries. The need for food has greatly grown due to the rise individual income, but agricultural output cannot keep up with the surge in demand. Theoretical justifications and time series econometric findings (Sasmal, 2015) demonstrate that rising individual income and a scarcity are to blame for price increases. The price of agriculture has no long-term link with the money supply.

It is imperative to increase food production (Carruthers et al., 1997) through irrigation, which will need expanding the irrigated area and water sources as well as increasing the efficiency with which existing water supplies are used. Both of these growth sources will be difficult, necessitating intricate institutional and policy changes. Failure to adequately expand and intensify irrigated agriculture leads to added strain on available land resource and speediness up the process of environmental degradation. Lack of knowledge about the connections between water shortage, agricultural productivity, food security, and environmental sustainability has impeded irrigation and water development plans. Research that would advance this knowledge would be highly profitable.

Although the Green Revolution helped India become self-sufficient, the country is currently rapidly nearing stability (Tiwari et al., 2023). Agricultural diversification is required high nutrients and sustainability. It is warning alarm for getting the food and nutrition. But getting there is really challenging. Data that is easily accessible indicates that there is a rise in world hunger. Tiny millets are wonderful crops that can help address some of the issues in this sector since they are nutrient-dense, low-labor, robust, and adaptable.

Tiny millets are taken far less frequently, according to the comparison of main grains in this study (Ridoutt & Garcia, 2020). Resilient, versatile, low-labor, nutrient-dense, and sustainable, tiny millets can help address some of the problems in modern agriculture. Compared to big grains, little millets are ingested far less frequently. The over-reliance on a few number of plants has led to the marginalization and neglect of many grains. Little millets may include complex carbohydrates, minerals, phytochemicals, and energy. Based on this studies, malnutrition, including under-nutrition and over-nutrition, may be effectively treated with these. Small-seeded grasses called millets appear to be a good fit. These offer a well-balanced macronutrient and micronutrient content and are safe for the environment. In India, tribes (Subramanyam & Narayana Rao, 2023) are regarded as one of the most marginalized and economically deprived populations, struggling with issues of poverty, marginalization, and deprivation on top of hunger and food insecurity. Tribal people's issues with hunger, poverty, and malnutrition are mostly related to the ecological and physical environment that they have been forced to adapt to since they have no other options for millennia.

Owing to future forecasts, the major challenge in 2030 and beyond (Gulati et al., 2023) is food. UN Population projected that the (2019), India will surpass China by 2027. The problem would be made worse by the fact that most of the Indians are reside in areas with declining natural resource quality and limited availability by 2030.

3 Research Gap

The major gap is the decision-makers strategically utilize their various and occasionally at-odds mandates and guiding principles in these talks. This research identifies the various gaps that are related to the income, pricing, and food regimes as a result of linked structural processes and customs that worsen unequal flow.

The first gap is to explain this discrepant development. To take into account this variance in the outcome, it is argued that key principles in the literature on the expansion of social rights should be changed. Higher income in particular is significant, but it is not a requirement to explain the expansion.

The second gap shows that high denomination notes made up roughly 36.5% of all notes as of March 2022. The RBI's 2020–21 annual report indicates an unexpected increase in the amount of money in circulation. The signs of compulsive buying (Lejoyeux & Weinstein, 2010; Muller et al., 2015) which might resemble yearning and withdrawal, are persistent, repeating purchases that become the main reaction to unpleasant situations and emotions.

Finally, to prevent biodiversity losses and other unfavorable changes to the earth system, farmland development should be kept to a minimum. Natural resources are finite. Food costs are predicted to remain steady or decline globally over the next three decades, but progress in decreasing hunger in underdeveloped nations will be gradual. Small declines in crop production growth would cause food costs to increase and hunger to get worse.

4 Conclusion

Indian Government has taken steps for transfers to help low-income households; more precise targeting might boost their effectiveness and lead to a more equitable allocation of income. In addition to financial and technological transfers from developed nations, regulatory attention must be paid to energy technology research and development as well as carbon pricing in order to set the Indian economy on a path towards low-carbon growth. Uneven pricing or monopolistic pricing regime of the same product, higher denomination of currency urges to compulsive buying i.e., high denomination notes made up roughly 36.5% of all notes as of March 2022. The RBI's 2020–21 annual report indicates an unexpected increase in the amount of money in circulation. Constant, recurring purchases that become the primary response to negative circumstances and emotions, as well as money that is not distributed evenly, are symptoms of compulsive purchasing, which might mimic withdrawal and desire. Rising public debts and the exchange rate have some impact on prices, even if they are not compelling. To explore possible food policies and determine the optimal course of action, assume that social welfare aims to stabilize prices and financial support for agri based peoples to encourage retaining these sector. Pradhan and Ghosh (2022) setting an appropriate carbon tax rate on coal can accomplish dual objectives: meeting emissions reduction goals while simultaneously generating economic benefits. Additionally, this taxation mechanism can accelerate the transition toward an economy with reduced carbon intensity.

References

Agrawal, T., & Agrawal, A. (2023). Beyond consumption expenditure: Income inequality and its sources in India. *Progress in Development Studies, 23*(1), 7–27. https://journals.sagepub.com/doi/abs/10.1177/14649934221101735

Asmita, A., & Kumar, A. (2023). Monopoly market. *EPRA International Journal of Multidisciplinary Research (IJMR), 9*(1), 189–190. https://www.eprajournals.net/index.php/IJMR/article/view/1405

Atkinson, A. B. (2016). How to spread the wealth: Practical policies for reducing inequality. *Foreign Affairs, 95*, 29. https://heinonline.org/HOL/LandingPage?handle=hein.journals/fora95&div=8&id=&page

Brown, Z. Y., & MacKay, A. (2021). *Competition in pricing algorithms* (No. w28860). National Bureau of Economic Research. https://www.nber.org/papers/w28860

Carruthers, I., Rosegrant, M. W., & Seckler, D. (1997). Irrigation and food security in the 21st century. *Irrigation and Drainage Systems, 11*, 83–101. https://link.springer.com/article/10.1023/A:1005751232728

Gouel, C., Gautam, M., & Martin, W. J. (2016). Managing food price volatility in a large open country: The case of wheat in India. *Oxford Economic Papers, 68*(3), 811–835. https://academic.oup.com/oep/article/68/3/811/1752195

Gouri, G. (2023). Basics of economics of competition and markets in product markets. In *A commissioner's primer to economics of competition law in India* (pp. 13–45). Springer Nature Singapore. https://link.springer.com/chapter/10.1007/978-981-19-9476-0_2

Gulati, A., Paroda, R., Puri, S., Narain, D., & Ghanwat, A. (2023). Food system in India. Challenges, performance and promise. In *Science and innovations for food systems transformation* (pp. 813–828). Springer International Publishing. https://library.oapen.org

Katoch, O. R. (2022). Determinants of malnutrition among children: A systematic review. *Nutrition, 96*, 111565. https://www.sciencedirect.com/science/article/abs/pii/S0899900721004275

Karthick, A., & Alamelu, K. (2024). Internet of things, artificial intelligence, big data and cloud application technology for acceleration on business resilience. In R. K. Misra, S. A. Purankar, D. Goel, S. Kapoor, & R. B. Sharma (Eds.), *Resilient businesses for sustainability*. Advanced Series in Management (Vol. 34A, pp. 153–160). Emerald Publishing Limited. https://www.emerald.com/insight/content/doi/10.1108/s1877-63612024000034a013/full/html

Lejoyeux, M., & Weinstein, A. (2010). Compulsive buying. *The American Journal of Drug and Alcohol Abuse, 36*(5), 248–253. https://www.tandfonline.com/doi/abs/10.3109/00952990.2010.493590

Lestage, R., & Kim, S. J. (2023). *Regulation of platform price discrimination on either or both sides of the market*. https://www.researchsquare.com/article/rs-2695184/v1

Muller, A., Mitchell, J. E., & de Zwaan, M. (2015). Compulsive buying. *The American Journal on Addictions, 24*(2), 132–137. https://onlinelibrary.wiley.com/doi/abs/10.1111/ajad.12111

Narayan, J., John, D., & Ramadas, N. (2019). Malnutrition in India: status and government initiatives. *Journal of Public Health Policy, 40*, 126–141. https://link.springer.com/article/10.1057/s41271-018-0149-5

Pradhan, B. K., & Ghosh, J. (2022). A computable general equilibrium (CGE) assessment of technological progress and carbon pricing in India's green energy transition via furthering its renewable capacity. *Energy Economics, 106*, 105788. https://www.sciencedirect.com/science/article/abs/pii/S0140988321006265

Ridoutt, B., & Garcia, J. N. (2020). Cropland footprints from the perspective of productive land scarcity, malnutrition-related health impacts and biodiversity loss. *Journal of Cleaner Production, 260*, 121150. https://www.sciencedirect.com/science/article/abs/pii/S0959652620311975

Sarkar, S., & Mehta, B. S. (2010). Income inequality in India: Pre-and post-reform periods. *Economic and Political Weekly*, 45–55. https://www.jstor.org/stable/25742070

Sasmal, J. (2015). Food price inflation in India: The growing economy with sluggish agriculture. *Journal of Economics, Finance and Administrative Science, 20*(38), 30–40. https://www.sciencedirect.com/science/article/pii/S2077188615000062

Schmalense, R. (1981). Monopolistic two-part pricing arrangements. *The Bell Journal of Economics*, 445–466. https://www.jstor.org/stable/3003566

Singh, M., & Kumar, A. (2023). A critical political economy perspective on Indian television: STAR, Hotstar, and live sports streaming. Triple C: Communication, capitalism & critique. *Open Access Journal for a Global Sustainable Information Society, 21*(1), 18–32. https://triple-c.at/index.php/tripleC/article/view/1395

Stern, N. (1987). The effects of taxation, price control and government contracts in oligopoly and monopolistic competition. *Journal of Public Economics, 32*(2), 133–158. https://www.sciencedirect.com/science/article/abs/pii/0047272787900090

Subramanyam, V., & Narayana Rao, S. (2023). Food insecurity, poverty and malnutrition among particularly vulnerable tribal groups in Andhra Pradesh: Role of the government to mitigate problems. *Indian Journal of Anthropological Research, 2*(1), 77–94. https://www.arfjournals.com/image/catalog/Journals%20Papers/IJAR/2023/No%201%20(2023)/6_Subramanyam.pdf

Tiwari, H., Naresh, R. K., Bhatt, R., Kumar, Y., Das, D., & Kataria, S. K. (2023). Underutilized nutrient rich millets: Challenges and solutions for India's food and nutritional security: A review. *International Journal of Plant & Soil Science, 35*(2), 45–56. https://journalijpss.com/index.php/IJPSS/article/view/2758

Customer Awareness and Factors Affecting CBDC Adoption in Visakhapatnam, Andhra Pradesh: A Study on RBI's Digital Rupee in India

Battula Vijay Kiran, Teki Yaswanth Kumar, and A. Narasimha Rao

Abstract This research delves into the level of understanding and the attitude of the customers towards the retail CBDC of the Reserve Bank of India in Visakhapatnam, Andhra Pradesh. The primary and secondary sources were used for data collection by means of a structured questionnaire which was distributed via WhatsApp and E-mail means. Out of 368 responses received, 360 were selected for analysis ensuring that the sampling method of convenience provided a wide range of variation. Descriptive statistics were used to investigate the demographic characteristics' distributions while the relationships between the variables were tested using Chi-Square tests. It was found out that 82.5% of the respondents do not know about CBDC and level of education has a significant impact with the degree of awareness. Some factors including age, gender, and occupation were however not significant. Although more people are embracing the use of digital payments, the use of CBDC was rather low attributable to security issues and low usage push from the banks. The study recommends the launching of education programs, increasing the active role of banks, and bettering the characteristics of the CBDC.

Keywords Central bank digital currency (CBDC) · Digital payments · Customer awareness · Reserve bank of India (RBI)

1 Introduction

Digital technologies have been evolving with new faces coming into world financial systems by introducing new forms of payments and settlements. One such innovation in that respect has emerged as CBDC. In December 2022, the Reserve Bank of India

B. V. Kiran (✉) · T. Y. Kumar · A. Narasimha Rao
DCMS, Andhra University, Visakhapatnam, India
e-mail: vijaykiranslp@gmail.com

T. Y. Kumar
e-mail: yaswantht.rs@andhrauniversity.edu.in

R. S. Mor et al. (eds.), *Sustainable Business Management, Innovation and Technology*, Sustainable Economy and Ecotechnology,
https://doi.org/10.1007/978-3-031-98141-8_17

decided to introduce a CBDC in the form of the digital rupee to facilitate money transactions that are speedier, safer, and far more widely available. A CBDC is a virtual version of fiat money which co-exists with banks of physical cash to substitute monetary systems, conventionally doing business, in digital forms at less cost. Giant diversity in the population builds unique problems, standing in ways of using digitized use inductively. Among several platforms, this nation has spotted stellar growth in its UPI regarding digital payments.

There are, however, a few big factors that will determine the successful execution of the CBDC: one is the level of understanding among the citizenry, another is tech enablement, and the last is political support in general. While there are assumed gains from the use of CBDC, including cost reduction in transactions, improved safety and security, and an increase in access to finance, it is observed that acceptance depends critically on user and financial broker comprehension and confidence.

The research befits the growing discussion on digital payment's future in India by analyzing the association of technology, finance, and public policy. In this paper, the study identifies the factors of awareness, trust-building, and smooth integration in the financial system—all essentially a collective effort that will be required for the establishment of CBDCs.

2 Review of Literature

CBDCs are digital instruments of payment issued by central banks linked to the currency and are possibly one such evolution of monetary systems. In the case of CBDC adoption, awareness, confidence, and infrastructural preparedness will be important. According to Group of Central Banks (2020), a CBDC is a digital means of payment denominated in a national unit of account, a direct liability of the central bank, widely available for retail and consumer use. The Committee on Payments and Market Infrastructures (CPMI, 2018) further elaborates that CBDC is a digital form of central bank money for use outside rather than traditional reserve money, while Kumhof & Noone (2021) present them as a form of currency that is distinct from the traditional reserves, and also not convertible to and from reserves, with some variants also being interest-bearing digital currencies.

Kiff et al. (2020) further extend the definition, identifying CBDCs as being digital claims on central banks, representatives of the sovereign currency that can be tailor-made for different economies. Adrian & Mancini-Griffoli (2019) define it to say, "a direct CBDC is a liability of the central bank in which customer acquisition and Know Your Customer, or KYC, checks, through payment processing, are taken care of by the central bank itself. This tends to disintermediate the banks". Auer et al. (2020) and Bordo & Levin (2017) refer to indirect CBDC when the commercial banks have accounts with the central bank in return for CBDC reserves. Indirect CBDC looks very similar to today's financial system but is guaranteed with CBDC. In hybrid CBDC, both models are combined in such a way that the central bank provides partial infrastructure but leaves the consumption interface to commercial banks.

In other terms, the main demands that the consumers emphasize in a CBDC—in both its direct and indirect models—are stressed, such as privacy and ease of use. The consultation by the ECB underlines these aspects, emphasizing accessibility and speed (ECB, 2020).

Consumer Awareness and Perception: Alexander & Durai (2024) indicated that for the adoption of digital currencies, public awareness and understanding are essential. Their research identified factors like ease of use, trust in the issuing authority, and perceived security as significant influences on consumer behavior.

Technology and Behavioral Intentions: Previous research by Akhtar et al. (2023) examines using Unified Payments Interface (UPI) and revealed that prior experience with electronic payment systems influences the adopters likelihood to accept CBDCs. These drivers were identified as enjoyment while using (hedonic motivation) and expectations about the quality of execution of the system.

Role of Banks and Education Marketing: Further emphasizing on that the banking sector is in view CBDC enabling role by Lal Suresh in 2022, which mandates that intermediaries—bank and financial institution —need to educate consumers on the account of benefits of CBDCs (Lal, 2022). Also, Mishra & Singh (2023) proposed that a strong collaboration between banks and policy makers may lead to faster awareness as well as recognition.

Underlying Demographic/Structural Issues: Research by Osakwe et al. (2024) hint at infrastructural compatibility and literacy divides standing as obstacles for CBDC adoption. These roadblocks require specific awareness campaigns directed at the different demographic groups involved.

Perceived Benefits and Concerns: Gupta et al. (2023) For India's digital rupee, pressed that trust partially mediates between how perceived risk impacts upon willingness-to-adopt. Of interest, factors such as perceived ease of use, innovativeness and inertia did not significantly predict trust. The findings highlight the importance of managing these risks & growing trust for enhancing CBDC acceptance emphasizing a sounding policy needed by lawmakers.

Global Experiences and Lessons for India: Exploring international practices, Peterson Ozili (2023) identified the importance of careful CBDC design to balance monetary policy, consumer adoption, and financial system stability. The study emphasized learning from pilot programs in other nations.

2.1 Central Bank Digital Currency (CBDC): Concept, Models and Types

Central Bank Digital Currency (CBDC is) a digital representation of a country's fiat currency issued and administered by the central bank. While cryptocurrencies, such as Bitcoin are decentralized and speculative CBDCs are centralized (national financial authorities governs and controls CBDCs) In other words the CBDC is intended primarily to make the existing financial system more efficient and secure e.g. replacing physical cash, but with less risk created by lack-of-trust and uniqueness attributes that central banks embody. Globally, as crypto-everything is about to happen, CBDCs are getting a lot of fanfare from governments and financial institutions showing interest to change payments along with advancing the frontiers of financial inclusion/economic policy.

To issue a CBDC, there is need to have a robust integrated legal, technological and operational ecosystem of CBDC. Issuing a digital currency requires the central bank to make sure they have legal backing for doing so. It will need a legal regime of some type, it needs regulation, recognizing the nature of CBDC in its place and relation to the whole payments system. After that comes the relevant technologies, only once they being developed to build technological backbone from the legal framework. Blockchain or Distributed Ledgers (DLT)—The central bank could leverage technologies to enable secure, speedy and transparent transactions. As the decentralized element of blockchain offers a strong way for tracking transactions and ensures their security.

The other aspect that is also of importance is the channel through which the CBDC will be distributed, and this may either be directly from the central bank or through a commercial bank as an intermediary. In the direct model, the central bank has the oversight of all the CBDC issuance, launch and management of the consumer wallets. Although this model provides the central bank with a high level of control over CBDCs, it can be associated with costs such as the need for investments in technology and setting up infrastructure for managing the system. The indirect model, on the other hand, enables the central bank to launch the currency to the commercial banks that are then responsible for its distribution and interaction with the customers. This method could also be efficient but creates a problem of scope since it constrains the central bank's ability to track transactions on an individual level. It is also possible to have a combination of both hence giving a hybrid model.

2.2 Benefits and Challenges of CBDC Implementation

Many positive impacts are expected from the launch of CBDCs. Financial inclusion in general, as one key benefit. Individuals that do not have a bank account and are unbanked or underbanked, for example a lot people in developing countries—and hence financial exclusion in many parts of the world can benefit from having access to CBDCs which could make their own digital payment method.

They may also reduce transaction costs since there are fewer intermediaries involved in the settling of balances, and traditional financial transactions fees will be lower.

One of the major benefits is the improved security and transparency of us that can be provided by the CBDCs. For digital currencies, each transaction is recorded on an open functionally secure ledger ensuring transparency and lower chance of fraud or money laundering. Digital currencies are more secure than traditional payment methods, which are usually intermediated in individual accounts that can become the object of cyber-attacks.

However, adoption of a CBDC introduces its own set of challenges as well. Cyber-security is a big worry because the digital pieces of CBDC are open to hackers, data breaches and hacking that will inevitably happen. Moreover, maintaining the right custody of proper trade-hygiene: and at no cost for consumers to lazy consent. CBDC must strike a balance between the user privacy and making sure that the transaction is traceable to avoid corruption. For CBDCs to be successful, they must link up with the current payment system seamlessly. The reception of CBDCs will need existing financial infrastructure "made to last" which gives them consistency their traditional currencies to easily transition into this mechanism alongside money and other digital payment-based.

The Reserve Bank of India (RBI) is actively involved in the development and implementation of the central bank digital currency (CBDC), referred to as the digital rupee, in both wholesale and retail sectors.

Wholesale Segment (e₹-W): Launched on November 1, 2022, the wholesale CBDC focuses on settling secondary market transactions in government securities. This initiative aims to enhance the efficiency of the inter-bank market by reducing transaction costs and minimizing settlement risks (Press Information Bureau 2022a).

Retail Segment (e₹-R): The retail CBDC pilot commenced in December 2022, targeting a broader user base, including the private sector, non-financial consumers, and businesses. By December 2023, the pilot achieved a milestone of 1 million daily transactions. However, subsequent reports indicate a decline in transaction volumes, attributed to reduced promotional efforts and limited organic demand (Press Information Bureau 2022b).

Recent Developments: In April 2024, the RBI proposed expanding the retail CBDC's accessibility by enabling non-bank payment system operators to offer CBDC wallets. This move aims to broaden user engagement and test the system's resilience across multiple channels. As of August 2024, the retail CBDC pilot includes over 5 million users and 16 participating banks. The RBI emphasizes a cautious approach, focusing on understanding the CBDC's impact on users, monetary policy, and the broader economy before a full-scale rollout. These initiatives reflect the RBI's commitment to integrating digital currency into India's financial system, with ongoing assessments to ensure stability and efficiency.

3 Research Interest and Methodology

Digital Rupee can disrupt the retail transactions, and it does so via banking app integration of different banks. It should be mentioned that this work is being done at an interface where technological developments as well digital finance services rise rapidly. For many people and professional bankers, they must have the knowledge and comprehension of CBDC in order for this to be made successful. Lack of knowledge about CBDC, in this case the central bank digital currency casts wider a wider cloak of implications whether deserved or not. This is because ignorance of digital rupee will slow down adoption curve rails that which can be used as currency. Hence it become a area of existing knowledge, what the literacy about currency is and the information sources as well as; knowledge gaps or misconception is/are. Moreover, it is equally vital to know how the banks are propagating the information about the introduced CBDC and it's use case. When these aspects are known, then strategies can be made to raise the knowledge and attitude of the population about the CBDC.

This paper will be focusing on a large city in Visakhapatnam (largely due to its diverse demographic profile), situated in Andhra Pradesh, India. This city is a perfect place to assess awareness and factors that influence on CBDC adoption. The results of this study would provide significant insights to a more general insight on how the CBDC is viewed and inform potential strategies to further its implementation and assent into the public financial activities.

4 Objectives of the Study and Hypothesis

Assessment of the awareness and channels of information regarding CBDCs among customers in Visakhapatnam, the role of bank employees regarding the same.

The study shall also seek to analyze the trends in customers' preference for and perception of digital means of payments, emphasizing CBDC and desirable improvements.

To provide feedback and suggestions to have better CBDC awareness in terms of adaptation based on the collected customer feedback.

5 Rationale

The hypotheses were derived from the objectives and supported by existing research on technology adoption and financial behavior. Studies indicate that demographic factors such as gender, age, education level, and occupation influence awareness and usage of new financial technologies, including CBDCs (Adrian & Mancini-Griffoli, 2019; Central bank digital currencies: foundational principles & core features: Technical report, 2020). Gender differences often arise due to varying access to digital resources and financial literacy (Kumhof & Noone, 2021). Age is a key factor, with younger individuals typically more familiar with digital technologies (Kiff et al., 2020). Education level impacts the ability to comprehend and adopt innovations (CPMI, 2018), while occupation affects exposure to information channels like banks (Auer et al., 2020). These factors align with the objective of understanding how awareness and usage vary across customer segments in Visakhapatnam, enabling targeted strategies for improving CBDC adoption (Bordo & Levin, 2017).

6 Alternative Hypotheses

$\mathbf{H_{1a}}$: There is a significant relation between gender and awareness of Central Bank Digital Currency (CBDC).
$\mathbf{H_{1b}}$: There is a significant relation between age and awareness of CBDC.
$\mathbf{H_{1c}}$: There is a significant relation between education level and awareness of CBDC.
$\mathbf{H_{1d}}$: There is a significant relation between occupation and awareness of CBDC.
$\mathbf{H_{1e}}$: There is a significant relation between gender and usage of CBDC.
$\mathbf{H_{1f}}$: There is a significant relation between age and usage of CBDC.
$\mathbf{H_{1g}}$: There is a significant relation between education level and usage of CBDC.
$\mathbf{H_{1h}}$: There is a significant relation between occupation and usage of CBDC.

Sample for the Study.

Given the large population of Visakhapatnam, the sample size is calculated as 360 using the formula for an infinite population:

$$n = \left(\left(Z^2.p.(1-p)\right)/e^2\right)$$

$$n = \left(\left(1.96^2.0.5.(1-0.5)\right)/0.05^2\right) \approx 384$$

n sample size
p Estimated proportion of an attribute that is present in the population (0.5 for maximum variability)
Z Z-Value (1.96 for a 95% confidence)
e margin of error (0.05).

For practicality 360 sample is selected.

7 Data Collection

Data was collected using both primary and secondary sources. A structured questionnaire was distributed through platforms such as WhatsApp and email, targeting a diverse group of customers in Visakhapatnam. Initially, 368 responses were received, of which 360 were deemed usable after filtering. The final analysis was based on this refined sample of 360 respondents.

Statistical Techniques.

This study employs both descriptive and inferential statistics to analyze survey data and assess CBDC awareness among respondents. Descriptive statistics summarize key features of the data, such as frequency distributions to show response counts and percentages to compare proportions across groups and questions. These statistics highlight overall patterns, including the proportion of respondents familiar with CBDC, digital payment usage, and demographic characteristics.

Additionally, the Chi-Square test of independence is used to examine significant relationships between categorical variables, such as CBDC awareness (Yes/No), age groups, and education levels. This test identifies whether observed differences in awareness and usage across demographic groups are statistically significant or due to chance. Building on the approach used by Gao et al., which applies chi-square tests for assessing associations during the COVID-19 period, the current study adopts a similar methodology (Gao et al., 2020). While advanced statistical techniques like Structural Equation Modelling (SEM) or multivariate regression are suited for complex analyses, this study focuses on simple associations between categorical variables, such as demographic factors (e.g., age, gender, education) and CBDC awareness/usage, making the chi-square test a reliable choice.

The Chi-Square test effectively meets this goal by testing independence and examining relationships between distinct categories. Moreover, advanced techniques often require larger sample sizes, more complex data structures, and assumptions like normality or multicollinearity that do not align with the data characteristics or objectives of this research. Given the focus on providing actionable insights into CBDC awareness and adoption, the selected methods are sufficient and appropriate for capturing the key relationships of interest.

8 Results and Discussions

The survey conducted among customers in Visakhapatnam reveals important insights into the awareness and usage of the Central Bank Digital Currency (CBDC), as well as the role of banks in promoting this digital payment method.

The survey conducted among customers in Visakhapatnam reveals important insights into the awareness and usage of the Central Bank Digital Currency (CBDC), as well as the role of banks in promoting this digital payment method.

The Table 1 provides a comprehensive breakdown of respondents across various demographic categories, presenting both frequency and percentage distributions for each sub-category.

In the **Age** category, respondents are fairly distributed, with the highest representation in the 46–55 age group (26.67%), followed by those aged 26–35 (17.22%) and 36–45 (16.39%). Respondents aged 18–25 and above 55 each account for 15.83%, while the below-18 group constitutes the smallest share at 8.06%.

For **Gender**, female respondents form the majority, representing 63.89% of the sample, while males account for 30.83%, and 5.28% identified as 'Other.'

In terms of **Education Level**, 26.39% of respondents have completed matriculation or below, followed closely by postgraduates at 25.83%. Both above matriculation and undergraduate categories each represent 22.5%, while only 2.78% hold a professional degree.

Table 1 Demographic descriptive statistics

Category	Sub-category	Frequency	Percentage (%)
Age	Below 18	29	8.06
	18–25	57	15.83
	26–35	62	17.22
	36–45	59	16.39
	46–55	96	26.67
	Above 55	57	15.83
	Total	360	100
Gender	Female	230	63.89
	Male	111	30.83
	Other	19	5.28
	Total	360	100
Education level	Matriculation or below	95	26.39
	Above matriculation	81	22.5
	Undergraduate	81	22.5
	Postgraduate	93	25.83
	Professional degree	10	2.78
	Total	360	100
Occupation	Student	95	26.39
	Self-employed	91	25.28
	Employed	74	20.56
	Unemployed	88	24.44
	Retired	12	3.33
	Total	360	100

Source Self-Calculated based on the survey responses

Lastly, the **Occupation** category reveals that students make up the largest group at 26.39%, closely followed by the self-employed (25.28%) and unemployed (24.44%). Employed individuals comprise 20.56%, and retirees form the smallest group at 3.33%.

This detailed demographic segmentation highlights the diversity of the respondents, ensuring a well-rounded representation for the analysis (Table 2).

8.1 Awareness and Usage of CBDC

A significant majority of respondents (82.00%) have not heard about CBDC, highlighting a substantial gap in awareness. Among the 18.00% who are aware, the sources of information are varied, with 29.00% learning about CBDC from the news, 17.00% from friends or family, 17.00% from banks, 17.00% from social media, and 19.00% from other sources.

Usage of CBDC remains very low, with 91.00% of respondents having never used it, compared to 9.00% who have. Additionally, knowledge of the benefits of using CBDC is also low, with only 8.00% aware of the benefits and 92.00% not aware. Concerns about security are prevalent, as 91.00% of respondents do not consider CBDC secure, compared to 9.00% who believe it is secure.

8.2 Digital Payment Methods

The adoption of digital payment methods such as UPI, cards, and mobile wallets is balanced, with 50.00% of respondents using them and 50.00% not using them. The frequency of digital payment usage compared to cash varies, with 22.00% using both equally, 22.00% mostly using cash, 21.00% mostly using digital methods, 19.00% always using digital methods, and 18.00% always using cash.

When comparing CBDC to other digital payment methods, opinions are mostly uncertain: 2.00% believe CBDC is better, 3.00% do not, and 94.00% are unsure. Respondents identified several areas for improvement in CBDC, including better security (3.00%), more user-friendly features (35.00%), wider acceptance (26.00%), more incentives (32.00%), and other improvements (3.00%).

8.3 Role of Banks in Promoting CBDC

It has not been reported by a most respondents (91.00 that bank personnel have informed them about CBDC, and only 9.00% indicated having been informed about this by their bank. In addition, 91.00% of subjects think that the banks are not putting enough effort to help spread CBDC, while 9.00% think that the banks are

Table 2 Summary statistics (%) of survey

Section	Question	Response	Percentage
Awareness and usage of CBDC	Have you heard about CBDC (digital rupee)?	Yes	17.5
		No	82.5
	If yes, how did you first hear about CBDC?	Bank	17.46032
		Social media	17.46032
		News	28.57143
		Friends/Family	17.46032
		Other	19.04762
	Have you used CBDC?	Yes	8.611111
		No	91.38889
	Do you know the benefits of using CBDC?	Yes	8.333333
		No	91.66667
	Do you think CBDC is secure?	Yes	8.888889
		No	91.11111
Digital Payment Methods	Do you use digital payment methods (e.g., UPI, cards, mobile wallets)?	Yes	50.27778
		No	49.72222
	How frequently do you use digital payments compared to cash?	Always digital	18.61111
		Mostly digital	20.55556
		Equally digital and cash	21.66667
		Mostly cash	39.16667
	In comparison to other digital payment methods, do you find CBDC better?	Yes	2.5
		No	3.055556
		Not sure	94.44444
	What improvements would you expect in CBDC?	Better security	3.225806
		More user-friendly	35.48387
		Wider acceptance	25.80645
		More incentives	32.25806
		Other	3.225806
Role of banks in promoting CBDC	Have bank employees informed you about CBDC?	Yes	9.166667
		No	90.83333
	Do you feel that banks are doing enough to promote CBDC?	Yes	8.888889
		No	91.11111

Source Self-Calculated based on the survey responses

doing enough on this topic. In this way, these results raise an urgent call for the banks to increase their actions for raising awareness and educating their customers on CBDC and to also to improve security, incentives and user-friendliness in order to spur adoption.

Banks play a key role in this, and their role in CBDC development should be more active and visible.

Results of the survey are numerically presented in Table 3 and they describe CBDC knowledge and uptake by different demographic categories such as gender, age, education, and employment status. It depicts the ratio of respondents saying they have ever heard of CBDC to those saying they have not, and the ratio of respondents who have and have not used CBDC. The data show variations in terms of awareness and usage patterns among these groups and provides important information for understanding the drivers behind CBDC adoption.

Table 3 Survey results on CBDC awareness and usage

Category	Not heard	Heard	Not used	Used
Gender				
Female	188	42	208	22
Male	93	18	102	9
Other	16	3	19	0
Age				
18–25	45	12	52	5
26–35	53	9	57	5
36–45	51	8	55	4
46–55	76	20	85	11
Above 55	46	11	52	5
Below 18	26	3	28	1
Education level				
Above matriculation	62	19	73	8
Matriculation or below	80	15	88	7
Postgraduate	80	13	87	6
Professional degree	5	5	7	3
Undergraduate	70	11	74	7
Occupation				
Employed	63	11	70	4
Retired	9	3	11	1
Self-employed	72	19	81	10
Student	77	18	86	9
Unemployed	76	12	81	7

Source Self-Calculated based on the survey responses

The Chi-Square Test of Independence is used to determine if there is a significant association between two categorical variables. Table 4 summarizes the results of Chi-Square tests for awareness and usage of CBDC across different demographic factors.

The results in Table 4 indicate that there is no significant association between the compared variables and awareness of CBDC, as all p-values are above the common significance threshold of 0.05. This suggests that awareness of CBDC is uniformly low across different demographic groups and that the efforts made by banks to inform and promote CBDC have not significantly impacted awareness except for education level (Table 5).

Table 4 Chi-Square test on data

CBDC Category	Variable	Chi-Square Statistic	p-value	Interpretation
Awareness	Gender	0.257	0.879	No significant association between gender and awareness of CBDC
	Age	3.410	0.637	No significant association between age and awareness of CBDC
	Education level	11.160	0.025	Significant association between education level and awareness of CBDC
	Occupation	2.591	0.628	No significant association between occupation and awareness of CBDC
Usage	Gender	2.092	0.351	No significant association between gender and usage of CBDC
	Age	2.250	0.814	No significant association between age and usage of CBDC
	Education level	6.716	0.152	No significant association between education level and usage of CBDC
	Occupation	1.759	0.780	No significant association between occupation and usage of CBDC

Source Calculated based on the survey responses using Microsoft Excel Data Analysis Tool

Table 5 Hypotheses testing

Hypothesis No	Alternative hypothesis	Accept/ Reject
H_{1a}	There is a significant relation between gender and awareness of CBDC	Reject
H_{1b}	There is a significant relation between age and awareness of CBDC	Reject
H_{1c}	There is a significant relation between education level and awareness of CBDC	Accept
H_{1d}	There is a significant relation between occupation and awareness of CBDC	Reject
H_{1e}	There is a significant relation between gender and usage of CBDC	Reject
H_{1f}	There is a significant relation between age and usage of CBDC	Reject
H_{1g}	There is a significant relation between education level and usage of CBDC	Reject
H_{1h}	There is a significant relation between occupation and usage of CBDC	Reject

Source Self constructed

8.4 Awareness of CBDC

Gender: Chi-Square test results show that there is no significant relationship between gender and CBDC awareness, as evidenced by Chi-Square statistic of 0.257 and p-value of 0.879. The degree to which respondents indicate awareness of the CBDC seems roughly equal between genders, suggesting that gender has a negligible impact on the level of awareness among respondents.

Age: Chi-Square test results for age also indicate no relationship with level of awareness of CBDC, Chi- Square statistic 3.410, and p-value 0.637. This implies that awareness of CBDC is relatively stable at different age groups, and that whether one knows CBDC is not significantly affected by age, i.e., an individual of any age is equally able to know the existence of CBDC, and age is not a critical factor in awareness.

Education Level: Results on education level show a large association with level of awareness of CBDC, as indicated by Chi-Square statistic (11.160) and p-value (0.025). This shows that awareness of CBDC varies significantly depending on education level. Individuals from different education backgrounds display varying degrees of awareness, which underlines the importance of educational efforts in promoting awareness of CBDC.

Occupation: Chi-Square test findings for occupation show no association with CBDC awareness, a Chi-square statistic of 2.591 with $p = 0.628$. This suggests that awareness of CBDC is quite homogenous across professional groups, in other words, the awareness of CBDC is not significantly related to professional activities.

8.5 *Usage of CBDC*

The Chi-Square Test results are that there is no statistically significant evidence of any gender effect on CBDC use, a Chi-Square statistic of 2.092 and a p-value of 0.351. This makes it plausible that there are no differences in the probability of CBDC use between genders and, hence, that gender has little impact on the use of CBDC. In terms of age, the Chi-Square test findings also find no significant relationship with CBDC ownership, with Chi-Square statistic 2.250 and a p-value of 0.814. This also suggests that CBDC usage among different age groups is not substantially different from each other, indicating that the age itself does not play a significant role to predict the usage of CBDC. With respect to the level of education, Chi-square test does not indicate any significant relationship to the use of CBDC, with a Chi-square statistic of 6.716 and p-value of 0.152.

This means that the probability of CBDC usage is not significantly different across education groups, that is, the probability of CBDC usage does not differ significantly with education. In case of occupation, the Chi-Square test actually reveals no statistically significant relationship with CBDC adoption, with a Chi-Square statistic of 1.759 and p-value of 0.780. This also implies that CBDC usage is consistent over different jobs, meaning that the job does not play an important role in whether or not individuals utilize CBDC.

In conclusion, education level is the only factor significantly associated with awareness of CBDC, suggesting that individuals with different educational backgrounds have varying levels of awareness.

However, gender, age, and occupation do not significantly impact awareness or usage of CBDC. This indicates that while educational initiatives may be crucial for enhancing awareness, other demographic factors like gender, age, and occupation may not require as much emphasis in promoting CBDC usage.

9 Conclusions and Suggestions

The survey aimed at evaluating customer awareness and usage of the retail Central Bank Digital Currency (CBDC) issued by the Reserve Bank of India (RBI) reveals several important insights. Awareness of CBDC among customers is significantly low, with 82.50% of respondents indicating they have not heard of it. This not only obscures the overall understanding but also highlights the urgent need for comprehensive educational promotional activities to inform the public about the practical roles and benefits of CBDC. A Chi-Square test indicates that education level greatly influences awareness; individuals with professional degrees are more likely to be aware of CBDC compared to those with lower educational backgrounds. In contrast, demographic factors such as age, gender, and occupation seem to have little to no impact on awareness levels, suggesting that the lack of awareness is consistent across various demographic groups. Regarding the usage of CBDC, the survey findings suggest that factors like age, education level, gender, and workplace do not significantly influence individuals' decisions to use CBDC. This indicates that when respondents are educated about CBDC, their age does not play a crucial role in their usage decisions. However, while there seems to be no discernible pattern or bias in the usage percentages, the overall adoption appears to be limited, indicating a small circulation of the currency. The role of banks in promoting CBDC has become a notable concern. A majority of respondents (91.11%) believe that banks are not sufficiently educating customers about CBDC, with only a small fraction (9.17%) reporting that they learned about CBDC from bank employees.

All this calls for banks to actively consider the role for CBDC by utilizing their existing client relationships and communication infrastructure. Institutes of apex bank's have to constitute targeted programmes especially concentrating on the people with less level of awareness. Financial intermediaries should be more involved in CBDC promotion, in providing as much information to all clients as possible with respect to the advantages of and appropriate utilization of CBDC, achieved through systematic staff training, and through customer-oriented outreach activities. Further, by taking steps to dismantle gender imbalances in CBDC use (i.e., by implementing inclusive measures), greater uptake and acceptance can be achieved. It is advisable, therefore, to conduct regular surveys and evaluations to follow awareness and usage trends and to evaluate the long-term effectiveness of promotional activities. Through the adoption of these approaches, RBI and banking institutions can increase awareness and acceptability of CBDC and play a major role in the overall aim of stimulating digital payments in India. These activities will also help the public to cope with the introduction of CBDC and any tech additions, thus facilitating the smooth integration of the financial system to a digitalized one without introducing any damaging effect on traditional, bank-based payment schemes.

References

Adrian, T., & Mancini-Griffoli, T. (2019). *The rise of digital money*. IMF FinTech Notes, No. 19/0018, Washington, D.C.: International Monetary Fund.

Akhtar, F., et al. (2023). Does previous experience with the unified payments interface (UPI) affect the usage of central bank digital currency (CBDC)? *Journal of Risk and Financial Management, 16*(6), 286.

Alexander, C. P., & Durai, T. (2024). Digital currency: An awareness of the future of central bank digital currency in India. *International Journal of Services, Economics and Management, 15*(6), 667–676.

Auer, R., Cornelli, G., & Frost, J. (2020). *Rise of the central bank digital currencies: Drivers, approaches, and technologies*.

Bordo, M., & Levin, A. (2017). *Central bank digital currency and the future of monetary policy (w23711)*. Cambridge, MA: National Bureau of Economic Research, Article w23711.

CPMI. (2018). *Central bank digital currencies: Technical report*.

ECB. (2020). *Report on a digital euro: Technical report*, (p. 55).

Gao, Z., Ying, S., Liu, J., Zhang, H., Li, J., & Ma, C. (2020). A cross-sectional study: Comparing the attitude and knowledge of medical and non-medical students toward 2019 novel coronavirus. *Journal of Infection and Public Health, 13*(10), 1419–1423.

Group of Central Banks. (2020). *Central bank digital currencies: Foundational principles and core features: Technical report.*

Gupta, S., Pandey, D. K., Ammari, A. E., & Sahu, G. P. (2023). Do perceived risks and benefits impact trust and willingness to adopt CBDCs? *Research in International Business and Finance 66.*

Kiff, J., Alwazir, J., Davidovic, S., Farias, A., Khan, A., Khiaonarong, T., Malaika, M., Monroe, H., Sugimoto, N., Tourpe, H., & Zhou, Z. (2020). A survey of research on retail central bank digital currency. *SSRN Electronic Journal.*

Kumhof, M., & Noone, C. (2021). Central bank digital currencies—design principles for financial stability. *Economic Analysis and Policy, 71*, 553–572.

Lal, S. B. (2022). Central bank digital currency in India—an overview. In Bhowmik, D. (Ed.), *An approach towards CBDC*. New Delhi: Kunal Books.

Mishra, A., & Singh, T. (2023). Public perceptions of CBDC in India. *IJFMR, 6*(3), 1–10.

Osakwe, C. N., Ogunmokun, O. A., Elgammal, I., & Kwarteng, M. A. (2024). Individuals' attitudes and their adoption intentions of central bank digital currency: Combining theories and analytics for deeper insights. *Technological Forecasting and Social Change, 210*, Article 123857.

Ozili, P. K. (2023). Central bank digital currency research around the world: A review of literature. *Journal of Money Laundering Control, 26*(2), 215–226.

Press Information Bureau. (2022a). *Central bank digital currency (CBDC) pilot launched by RBI in retail segment has components based on blockchain technology*. Government of India.

Press Information Bureau. (2022b). *Central bank digital currency (CBDC): e₹-R is in the form of a digital token that represents legal tender*. Government of India.

Data Sovereignty and Cross-Border Privacy Laws in the Age of AI: India Versus Global Standards

Partho Banerjee and Gagandeep Kaur

Abstract This research study offers a comprehensive review of data privacy and protection legislation across key foreign jurisdictions, specifically India, the European Union (EU), the United Kingdom, the United States, Australia, and Japan. It analyses the distinct attributes, advantages, and obstacles of each framework, with the objective of uncovering insights to improve worldwide data protection standards. **The Digital Personal Data Protection (DPDP) Act of 2023** and the **Information Technology (Intermediary Guidelines and Digital Media Ethics Code) Rules of 2021** in India underscore consent-driven data processing and regulatory supervision. Nonetheless, the execution of these steps poses considerable operational difficulties. The EU's General Data Protection Regulation (GDPR) is acknowledged as a rigorous and thorough data protection framework with tough compliance requirements; however, it draws criticism for elevated compliance costs and uneven enforcement among member states. The United States does not possess a comprehensive federal privacy legislation, opting instead for a sector-specific framework exemplified by statutes like the California Consumer Privacy Act (CCPA) and the Health Insurance Portability and Accountability Act (HIPAA). This disjointed methodology leads to regulatory discrepancies and intricate compliance environments. Australia and the United Kingdom, although affected by the GDPR, have developed customised frameworks that reflect local particularities, whereas Japan's Personal Information Protection Law (PIPL) exhibits a more state-centric methodology with stringent data localisation mandates. With the advancement of digital technology, the safeguarding of data privacy and security is becoming progressively vital. This article delineates the principal similarities and variations across these legal regimes, emphasising the necessity for enhanced international regulatory cooperation and uniform data protection standards to more effectively safeguard personal data globally.

P. Banerjee (✉)
Advocate at High Court of Judicature Allahabad, Prayagraj, Uttar Pradesh, India
e-mail: partho.banerjee07@outlook.com

G. Kaur
Department of Law, University of Petroleum and Energy Studies, Dehradun, India
e-mail: gkaur@ddn.upes.ac.in; gdkaur17@gmail.com

R. S. Mor et al. (eds.), *Sustainable Business Management, Innovation and Technology*, Sustainable Economy and Ecotechnology,
https://doi.org/10.1007/978-3-031-98141-8_18

Keywords Data protection · Data privacy · General data protection regulation (GDPR) · European union (EU)

1 Introduction

The concept of privacy has been vital throughout history, with early thinkers such as Aristotle and John Locke distinguishing between public and private areas. Westin pioneered modern privacy theory, which defines privacy as the right to regulate how personal information is disclosed. Nonetheless, the emergence of the internet and digital technology has posed significant challenges to conventional concepts of privacy. Data privacy, or control over personal data in digital contexts, has become a major problem, particularly as advances in technology heighten the tension between privacy rights and the massive data collecting required for the digital economy.

The progress of science and technology has significantly changed global lifestyles, with advancements like the telephone and medicinal drugs serving as historic milestones. Among them, the internet has been a crucial engine of change from its inception in the late 1950s, affecting both fundamental needs and luxury while also increasing global connectivity. This interconnection is enabled by the power of the internet and smartphones, which have transformed commerce and education, decreasing the world into a tiny, accessible marketplace and allowing for online learning from anywhere.

In past centuries, oil was regarded as the world's most precious resource, pushing economies forward. In the contemporary digital era, data has surpassed oil in value (The Economist., 2017). By January 2021, 4.32 billion individuals will be connected to the internet, making data the new currency of the global economy. The COVID-19 (Dwivedi et al., 2020) pandemic emphasised the significance of data protection, as millions of people migrated to remote work, forcing traditional firms to integrate internet operations. The increase in digital data, with 90% created in the last decade, has highlighted the necessity for strong data protection. This is especially important when data privacy concerns rise, with consumers exchanging personal information for the convenience of internet services. The erosion of personal space in the digital age has driven the governments of all the countries to implement rules that safeguard citizens' data privacy. In order to preserve consumer trust, effective data protection strategies increasingly prioritise data restoration, preventing compromise, and maintaining privacy.

As data privacy becomes more crucial globally, several jurisdictions established legislative frameworks that reflect their legal traditions, socioeconomic conditions, and cultural values. This disparity complicates international data flows and worldwide company operations, necessitating the harmonisation of global standards and the enhancement of individual privacy rights through establishing an international body to regulate data privacy and protection akin to the International Telecommunication Union, which regulates the telecom industry globally.

2 Literature Review

Data protection regulations have changed significantly, particularly in developing nations such as India, providing valuable insights into the implementation of ***the Digital Personal Data Protection or (DPDP) Act, 2023*** & the **Information Technology (Intermediary Guidelines and Digital Media Ethics Code) Rules, 2021**. The primary IT Act **in force in India is the Information Technology Act, 2000**. This Act provides the legal framework for electronic governance, digital transactions, and cybercrime regulation.

The **Information Technology (Intermediary Guidelines and Digital Media Ethics Code) Rules, 2021** are a set of rules formulated under **the IT Act, specifically under its Sect. 87(2)**. These rules were introduced to address the growing concerns related to the digital ecosystem, including social media platforms, online news, and digital media. Globally, the European Union's General Data Protection Regulation (GDPR) has established a standard, impacting countries such as India in the development of their national data protection regulations. However, research indicates that, while GDPR is a robust model, its approach may not adequately reflect India's unique digital infrastructure and socioeconomic diversity.

India's efforts to build data privacy legislation encounter challenges due to its enormous population, increasing digital usage, and diverse levels of digital competence. Studies indicate that although the DPDP Act, 2023, and the IT rule, 2021, represent significant advancements, they still require enhancement to align with the unique requirements of the country. Comparative evaluations demonstrate that, while India's data protection rules accord with worldwide best practices in areas such as consent mechanisms and data minimization, they still fall short in terms of enforcement and data subject rights when compared to frameworks such as GDPR (Czarnocki et al., 2021).

A frequently debated subject is the challenges of enforcing these regulations in a diversified society. Businesses face operational challenges due to ambiguous legislation language, which necessitates clearer standards and increased stakeholder interaction (Singh et al., 2023). The literature also discusses the impact of developing technologies such as AI and blockchain, which, while improving data security, pose a new regulatory difficulty.

India's approach to data protection has worldwide consequences, influencing its position in international data governance. Compliance with global norms, such as GDPR, is critical, but India must also examine its geopolitical and economic environment. The literature represents multiple stakeholder perspectives, with corporations concerned about compliance constraints and civil society advocating for strong privacy rights. Policy recommendations in the literature emphasize the need for improved legislative clarity, easier compliance, and increased stakeholder participation. For India's data protection rules to be effective, they must adapt to technological advances while protecting privacy rights.

In the end, the literature offers a comprehensive understanding of the challenges and prospects within India's data protection system, underscoring the necessity for continual enhancement and adjustment to address the requirements of a digital society.

2.1 Data Privacy in the Modern World and its Importance on Individual's Life

Innovations in technology have exacerbated the conflict between privacy rights and the massive data collection of users that powers the digital economy. AI and new storage technologies are allowing both corporations and governments to acquire more precise personal information (Fahey., 2022). This has underscored privacy concerns, stressing the necessity for individuals to possess control over their personal data, to provide informed consent, and to retain the ability to remove or transfer data.

The widespread use of digital gadgets and surveillance has psychological consequences, altering traditional notions of the privacy. Furthermore, the misuse of personal data, as demonstrated in the Cambridge Analytica scandal, jeopardizes individual autonomy and democracy. To overcome these issues, privacy must become a common defence against manipulation. Legal frameworks must adapt by seeing personal data as a valuable asset and regulating its usage to protect democratic processes and human liberties Bennett and (Raab., 2020).

2.1.1 Data Privacy and Protection Frameworks Exists in India

The advent of digital technologies and the internet rendered data privacy a negligible worry in India. The 2017 landmark judgement of Supreme Court decision in ***Justice K.S. Puttaswamy (Retd.) v. Union of India*** established privacy as a fundamental right, paving the way for stronger data protection measures. Early efforts included the Information Technology (Reasonable Security standards and Procedures and Sensitive Personal Data or Information) Rules of 2021, which established fundamental principles such as consent, security standards, and privacy rules. However, they weren't broad in scope.

2.1.2 The New IT Rule, 2021

The Information Technology (Intermediary Guidelines and Digital Media Ethics Code) Rule of 2021 amended the IT rule of 2000 by imposing new responsibilities on intermediaries, such as content moderation and grievance redressal, and introducing digital media regulation to improve data privacy and user protection.

2.1.3 Digital Personal Data Protection Act, 2023

The DPDP Act of 2023 is a big step forward in India's data protection framework, harmonising with worldwide norms such as the GDPR. (Sundara & Narendran., 2023) The key provisions include:

I. *Consent and Transparency*: The law requires explicit consent for data collection, along with detailed transparency requirements (Motiwala., 2024).
II. *Data Protection Authority (DPA)*: Establishes a DPA to oversee compliance, address grievances, and enforce regulations (Bashir and Nisar., 2020).
III. *Individual Rights*: These rights confer upon persons the capacity to access, rectify, erase, and transfer their personal data, thereby enhancing their control over their information.
IV. *Roles and Responsibilities*: Defines the roles of data fiduciaries and processors, emphasizing data security and compliance.

2.1.4 Implementation Complexity

Challenges include the cost and resource allocation required for compliance, sector-specific adaptations, operational impacts due to data localization, and complications in international data transfers. The effectiveness of the DPA in enforcing regulations, addressing grievances, and imposing penalties will be crucial for successful implementation.

2.1.5 Legal Implications

The DPDP Act is subject to ongoing revisions based on feedback from stakeholders, aiming to refine the legislative framework. India's evolving data privacy laws are expected to influence global norms and enhance trust in its digital economy. The balance between privacy and innovation, integration with global standards, and enhancing digital literacy are key considerations for fostering a healthy digital ecosystem European Union (EU) (2016).

3 European Union (EU) Data Protection Framework

The European Union always takes the lead in data protection and privacy legislation. The journey commenced with the Data Protection Directive 95/46/EC, instituted in 1995, which sought to safeguard individuals' personal data and facilitate the unrestricted flow of data across the EU. The Directive defined essential principles for data processing and mandated Member States to enact national legislation consistent with its provisions. However, significant technical improvements and the emergence of the digital economy needed a more comprehensive and standard approach, which resulted in the adoption of the General Data Protection Regulation (GDPR) (Ivanova., 2021) (Fig. 1).

The GDPR, effective May 25, 2018, is a significant overhaul of the EU's data protection framework. It seeks to unify data privacy regulations throughout the EU, provide individuals with enhanced safeguards and rights, and guarantee more rigorous enforcement. The GDPR has created an international standard for data protection, influencing legislation in other countries globally.

The GDPR has an extensive reach, covering any organisation that processes the personal data of EU people, irrespective of its geographical location. This extraterritorial reach means that businesses outside the EU are subject to GDPR if they handle EU residents' data, highlighting the EU's comprehensive approach to data protection (Choi., 2020).

The primary objective of the GDPR is to seek unambiguous, informed consent from data subjects before processing their information. Organizations must offer clear, transparent information regarding data collection, processing purposes, and individual rights. The grant of consent should be free, specific, informed, and unequivocal, with the option to withdraw at any time (Lynskey., 2018).

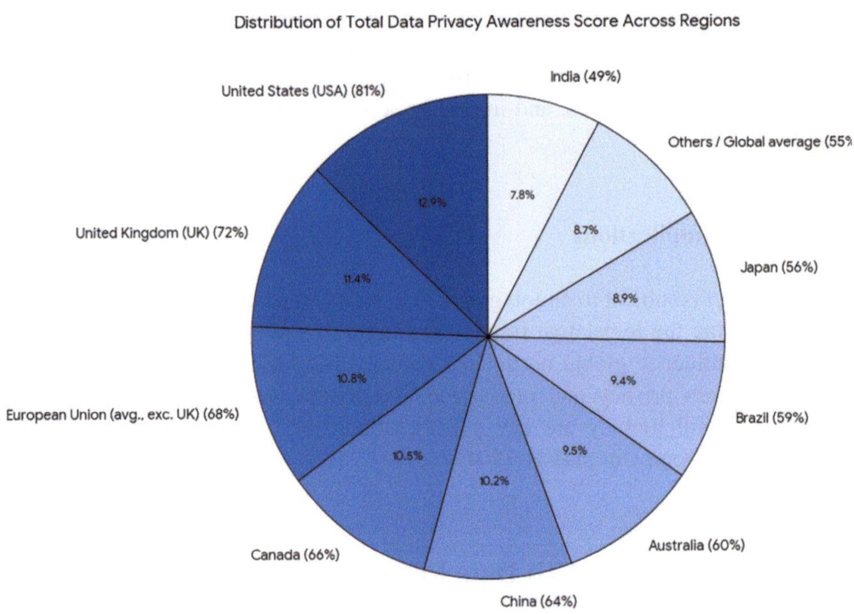

Fig. 1 Distribution of Data privacy and Policy Awareness. *Source Wikipedia &* and arxiv.org (Cornell University)

3.1 Data Subject Rights

The GDPR enhances individuals' rights over their personal data, encompassing:

- *Right to Access*: Individuals own the entitlement to request access to their data and understand its processing.
- *Right to Rectification*: Solicit amendments for inaccurate or incomplete information.
- *The Right to Erasure, or the right to be forgotten*, permits individuals to request the deletion of personal data under certain conditions.
- *Right to Restrict Processing*: Individuals possess the right to request limitations on data processing under certain conditions.
- *Right to Data Portability*: Permits individuals to transfer their data between service providers.
- *Right to Object*: Individuals may contest processing based on legitimate interests, direct marketing, or research/statistical objectives.

3.2 Enforcement and Penalties

The GDPR contains stringent enforcement procedures and significant fines for noncompliance (Kuner et al., 2021). Regulatory bodies in each EU Member State possess the authority to examine complaints, conduct audits, and impose fines of up to €20 million or 4% of global annual revenue, whichever amount is higher. This fosters accountability and adherence.

3.3 Challenges and Criticisms

Notwithstanding its advantages, the GDPR encounters obstacles:

- *Compliance Costs & Complexity*: The intricacy of the regulation incurs substantial compliance expenses, particularly for small and medium-sized enterprises (SMEs).
- *Interpretative Ambiguities*: Some provisions are open to interpretation, leading to inconsistencies across Member States.
- *Balancing Privacy and Innovation*: Maintaining strong data protection while fostering digital innovation remains challenging.
- *Global Impact and Jurisdictional Conflicts*: The GDPR's extraterritorial reach has caused conflicts with other jurisdictions' data protection laws, complicating international data transfers.

Recent EU data protection developments include the Privacy Regulation, which complements the GDPR by focusing on electronic communications privacy. Current legal disputes, including the annulment of the EU-US Privacy Shield, highlight the evolving landscape of EU data protection (Li et al., 2022). The GDPR continues to influence global data privacy standards, with the EU striving to balance data protection with the needs of the digital economy in the evolving landscape.

4 United States Data Privacy and Protection Framework

The United States' data privacy and protection framework has evolved incrementally, adopting a sectoral approach rather than a comprehensive national law. Early legislation, like the *Privacy Act of 1974*, laid the groundwork for federal data protection, focusing on how federal agencies handle personal data. However, the US approach has traditionally emphasised balancing privacy with business interests and technological innovation, leading to a fragmented system where various laws address specific sectors or types of data. The push for more comprehensive regulation has gained momentum due to growing public concern over data breaches, surveillance, and the widespread collection of personal information (Bamberger & Mulligan., 2011).

4.1 Key Legislation

- ***California Consumer Privacy Act (CCPA)***: The CCPA was enacted in 2018 and becomes effective on January 1, 2020. This significant state-level privacy regulation grants California residents' extensive authority over their personal data (California Legislative Information., 2018). These rights encompass understanding the gathered information, gaining access to and deleting it, and choosing not to sell personal data. The CCPA applies to enterprises that meet certain revenue or data-gathering levels, signalling a shift towards stricter consumer privacy safeguards at the state level.
- ***Health Insurance Portability and Accountability Act (HIPAA)***: In 1996, this significant federal legislation was enacted, regulating the privacy and security of health information. It defines national rules for the protection of sensitive patient information and governs the management of health data by healthcare providers, insurers, and their affiliates. The act also emphasises healthcare data privacy and security, which follows the industry-specific approach commonly followed in US privacy law.

4.2 Key Provisions

Consumer Rights and Protections: The CCPA affords consumers several key rights.

Consumers possess the right to comprehend the categories of personal data collected, its sources, and its applications and dissemination.

- *Right to Access*: Consumers are entitled to obtain their personal information.
- *Right to Erasure*: Consumers are entitled to seek the removal of their personal data.
- *Right to Opt-Out*: Customers possess the right to decline the sale of their personal information.

The CCPA prohibits businesses from discriminating against consumers exercising their rights.

Sector-Specific Regulations: US data privacy regulations are often sector-specific.

- *HIPAA (Health Insurance Portability and Accountability Act)*: Governs the privacy and security of health information.
- *Gramm-Leach-Bliley Act (GLBA)*: Regulates the confidentiality of financial data maintained by financial firms. The Children's internet Privacy Protection Act (COPPA) safeguards the internet privacy of children under the age of 13. Regulatory Instruments: The enforcement tactics differ based on the specific statute and jurisdiction.
- *CCPA*: Enforced by the California Attorney General, imposing penalties for non-compliance. It permits private litigation for specific data breaches. HIPAA is enforced by the Office for Civil Rights (OCR) of the Department of Health and Human Services (HHS), which possesses the jurisdiction to impose civil penalties.

4.3 Challenges and Criticisms

- *Fragmentation*: The absence of a comprehensive federal privacy law results in a fragmented array of state- and sector-specific requirements, complicating compliance for firms operating across multiple jurisdictions (Monahan., 2016).
- *Consumer Awareness*: Limited understanding of privacy rights among consumers can weaken the effectiveness of privacy laws.
- *Balancing Privacy and Innovation*: Finding a balance between privacy protections and fostering innovation remains a key challenge.
- *Enforcement Inconsistencies*: Variations in enforcement across states and agencies lead to inconsistent protection levels (Reis et al.,2024).

The CCPA serves as the model for new state-level privacy legislation like the Virginia Consumer Data Protection Act (VCDPA) and Colorado Privacy Act (CPA). There are ongoing debates about enacting a uniform federal privacy law to provide consistency across the U.S. The future of data privacy in the country points to more extensive and consistent legislation, which would be driven by public concerns and technical improvements. As the landscape changes, there is likely to be a renewed push for government privacy legislation that balances private rights with corporate requirements and innovation (Enright., 2022).

5 United Kingdom

- *Regulation*: The Data Protection Act of 2018 integrates the EU GDPR.
- *Enforcement*: Information Commissioner's Office.
- *Focus*: Data subjects' rights, special category data, data protection fees, offences, kid consent, and enforcement.
- *Brexit Impact*: Despite leaving the EU on March 29, 2019, the UK has maintained its data privacy laws, mirroring the GDPR (Diker Vanberg and Maunick., 2018).

6 Australia

- *Regulation*: Privacy Act 1988, including the Australian Privacy Principles (APPs).
- *Scope*: Applies to businesses with a turnover of $3 million or more.
- *Focus*: Consent, fairness, data breach notification, and access rights.
- *New Development*: The Consumer Data Right (CDR) empowers consumers and organisations to access and transfer their data across several sectors, akin to the portable data rights established by GDPR (Burdon & Mackie., 2020).

7 Canada

- *Regulation*: The Personal Information Protection and Electronic Documents Act (PIPEDA) (Austin., 2006).
- *Focus*: Regulates how personal information is collected, stored, and used online.
- *Requirement*: Privacy policies must be readily available to customers.

8 Japan

- *Regulation*: Personal Information Protection Act (PIPA).
- *Focus*: Broad definition of personal data, third-party consent, data transfers, record-keeping, anonymity, and breach handling (Wells & Ito., 2020).
- *EU Relationship*: Japan maintains a "white list" of EU countries that have implemented adequate data protection measures (Table 1).

Table 1 Comparative analysis of data protection law in India & major international jurisdictions

Aspect	India	USA	European union	UK	Brazil	Australia	Canada	China	Japan
Primary law	DPDP Act, 2023 New IT rule, 2021	Various federal and state laws (e.g., CCPA, HIPAA)	General data protection regulation (GDPR)	Data Protection Act, 2018	General Data Protection Law (LGPD)	Privacy Act 1988; Consumer Data Right (CDR)	Personal Information Protection and Electronic Documents Act (PIPEDA)	Privacy bill (2018)	Personal Information Protection Act (PIPA)
Scope	Comprehensive (once enacted)	Sectoral and state-specific	Comprehensive and uniform	Comprehensive, incorporating GDPR	Comprehensive and similar to GDPR	Focused on businesses over $3 M turnover	Comprehensive	Comprehensive, similar to GDPR	Broad definition; includes public directories
Data subject rights	Right to access, correction, data portability	Varies by state; e.g., CCPA provides access and deletion rights	Extensive rights: access, rectification, erasure, data portability	Extensive rights, mirroring GDPR	Similar rights to GDPR	Access, correction, and deletion rights	Access, correction, and deletion rights	Access and control over personal data	Broad rights, including third-party consent
Consent requirements	Explicit consent required	Varies; generally, requires opt-in for certain data	Explicit and informed consent required	Explicit and informed consent required	Explicit and informed consent required	Explicit consent for handling personal data	Explicit consent required	Explicit consent required	Explicit consent required
Data breach notification	Not yet specified (will be in the forthcoming law)	Varies by state; e.g., California requires notification	72 h notification requirement	72 h notification requirement	Notification requirement	Notification required within 30 days	Notification required	Notification requirement	Notification requirement

(continued)

Table 1 (continued)

Aspect	India	USA	European union	UK	Brazil	Australia	Canada	China	Japan
Children's data	Specific provisions expected in new law	Varies; e.g., COPPA for children under 13	Parental consent required for children under 16	Parental consent required for children under 16	Parental consent required for children under 16	Parental consent required for children under 16	Parental consent required for children under 13	Parental consent required for children under 16	Parental consent required for children under 16
Data protection authority	Pending (to be established under new law)	No central authority; varies by state and sector	European Data Protection Board (EDPB)	Information Commissioner's Office (ICO)	National Data Protection Authority (ANPD)	Office of the Australian Information Commissioner (OAIC)	Office of the Privacy Commissioner of Canada	National Information Security Standardization Technical Committee (NISSTC)	Personal Information Protection Commission (PIPC)
Cross-border data transfers	Governed by new law provisions	Varies; generally, less restrictive	Requires adequate protection or EU-approved mechanisms	Requires adequate protection or UK-approved mechanisms	Requires adequate protection or Brazil-approved mechanisms	Requires adequate protection or Australia-approved mechanisms	Requires adequate protection or Canada-approved mechanisms	Requires adequate protection or China-approved mechanisms	Requires adequate protection or Japan-approved mechanisms

9 Suggestions and Recommendations

In an era where data drives global economies and influences every aspect of daily life, strong personal data privacy is crucial. As technology advances and data flows across the borders at new speeds, the need for a uniform, international approach to data protection has never been higher. To fulfil this obligation, we propose establishing the **International Data Protection Authority (IDPA)**, an UN-based entity dedicated to establishing, maintaining, and harmonising worldwide data protection standards. The construction of a suggested worldwide data protection authority, akin to the International Telecommunication Union (ITU), necessitates the completion of several crucial steps to facilitate effective global coordination and comprehensive control.

The IDPA will be the central autonomous organisation responsible for developing comprehensive data protection frameworks that are consistent with various legislative norms while supporting universal privacy and security standards. Its objective will be to promote international collaboration, ease cross-border data transfers, and ensure that data protection procedures are effective and adaptive to the ever-changing digital environment. Below are some policy recommendations and suggestions:

I. ***Global Data Protection Framework***: Develop a unified framework with core principles inspired by the GDPR, ensuring uniformity while accommodating regional differences. The framework should emphasise transparency, data minimisation, purpose limitation, data subject rights, and security measures.
II. ***Rights of Data Subjects***: Guarantee that individuals possess the right to access, rectify, and transfer their data. We should need explicit, informed consent for data collection and processing, as well as transparent procedures for consent revocation.
III. ***Transparency and Timeliness***: Require regulators and affected individuals to report data breaches within a specified timeframe (e.g., 72 h), with transparency in breach notifications, including the nature of the breach and mitigation measures.
IV. ***Protection of Children's Data***: Implement stricter requirements for parental consent for processing data of individuals under a specified age (e.g., 16), and provide tools for parents to manage and monitor their children's data.
V. ***Cross-Border Data Transfers***: Establish mechanisms for recognising countries with adequate data protection standards, facilitating safe data transfers through standard contractual clauses and binding agreements.
VI. ***Data Protection Impact Assessments (DPIAs)***: Require DPIAs for processing operations that have a major influence on people's privacy in order to maintain openness and accountability.
VII. ***Enforcement and Dispute Resolution***: Develop robust enforcement mechanisms with the authority to impose significant fines for non-compliance and provide clear processes for resolving international data protection disputes.
VIII. ***Global Regulatory Body***: Establish an international authority, similar to the ITU, to oversee global data protection standards. This body would be

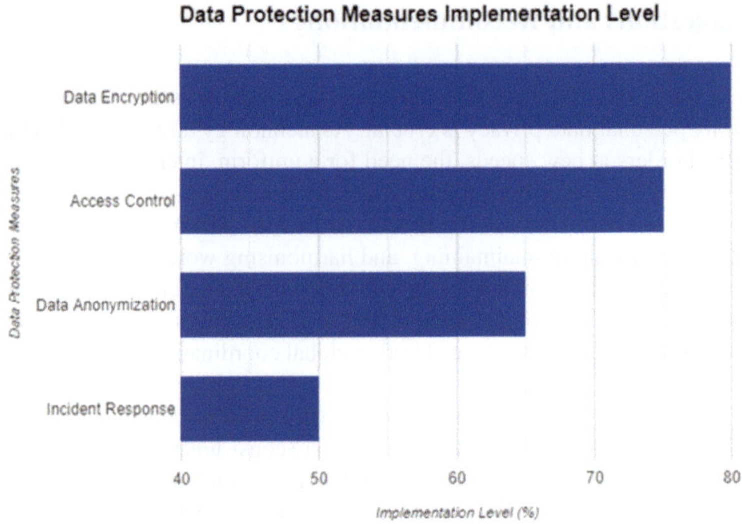

Fig. 2 Data protection measures implementation level. *Source* stationx.net

responsible for developing guidelines, monitoring compliance, and addressing cross-border data protection issues while fostering regional cooperation and information sharing.

IX. ***Technical support and capacity building***: Provide technical support, training courses, and advice to nations in order to help them adopt and enforce data protection regulations. Encourage international public awareness initiatives about the rights and obligations related to data protection.

X. ***Sustainable Funding and Support***: Develop a sustainable funding model to support the global body's activities, with contributions from member countries. Offer financial and technical assistance to developing nations to facilitate the implementation and enforcement of data protection policies (Fig. 2).

10 Conclusion

The rapid and profound digital transformation has fundamentally changed society, transforming data into a valuable yet challenging asset to safeguard, similar to the delicate balance needed to manage pollution in the face of industrial growth. As India's digital population surges, industries across healthcare, finance, e-commerce, and beyond are reimagining their services, increasingly collecting and utilising data to stay competitive. Yet, with this data boom comes an urgent need to safeguard individual privacy and secure sensitive information, as users unknowingly leave digital trails that could expose them to identity theft, financial crimes, and privacy breaches.

The crux of today's digital ecosystem lies in constructing a privacy policy that deftly balances individual rights with business needs, and this challenge calls for privacy policies to evolve beyond mere legal formalities. Businesses must craft privacy policies and terms of use as nuanced, adaptable frameworks that genuinely respect user consent and transparency. While the "notice and choice" mechanism remains foundational, adapting it to the dynamic and borderless nature of data flows is paramount.

The necessity for a unified global strategy for data protection has never been more evident. The proposed International Data Protection Authority (IDPA) represents a significant advancement in unifying various legislative frameworks, harmonising worldwide data protection standards, and promoting a unified approach to data security. The IDPA's potential to standardise privacy protections, provide technical assistance, and promote transparency would significantly enhance trust in cross-border data governance, ensuring that individual rights are upheld regardless of where data is processed or stored.

The establishment of a worldwide regulatory entity like the IDPA signifies not merely a regulatory achievement but a dedication to a digital future that harmonises innovation, economic advancement, and privacy rights. As the global digital landscape evolves, the IDPA's role in aligning distinct legal systems and fostering international collaboration may be pivotal in establishing a resilient, fair, and transparent digital economy, one that respects individual privacy as foundational to technological progress and economic expansion. Through these concerted efforts, a global data protection standard could be the cornerstone of a safer, more equitable, and privacy-centric digital world.

References

Austin, L. M. (2006). Reviewing pipeda: Control, privacy and the limits of fair information practices. *Canadian Bussiness Law Jounal, 44*, 21.

Bamberger, K. A., & Mulligan, D. K. (2011). *Privacy on the ground: Driving corporate behaviour in the United States and Europe*. Cornell University Press.

Bashir, M., & Nisar, M. A. (2020). Expectation versus reality: Political expediency and implementation of right to information laws. *Public Administration Quarterly, 44*(1), 3–30.

Bennett, C. J., & Raab, C. D. (2020). Revisiting the governance of privacy: Contemporary policy instruments in global perspective. *Regulation & Governance, 14*(3), 447–464.

Bhattacharjee, R. (2024). Data protection for democratic E-governance in India. *Indian Journal of Public Administration, 70*(3), 631–637.

Burdon, M., & Mackie, T. (2020). Australia's consumer data right and the uncertain role of information privacy law. *International Data Privacy Law, 10*(3), 222–235.

California Legislative Information. (2018). *California Consumer Privacy Act (CCPA)*. Retrieved from https://leginfo.legislature.ca.gov/faces/billTextClient.xhtml?bill_id=201720180AB375. Last accessed on December 20, 2024

Caron, X., Bosua, R., Maynard, S. B., & Ahmad, A. (2016). The internet of things (IoT) and its impact on individual privacy: An Australian perspective. *Computer Law & Security Review, 32*(1), 4–15.

Choi, K. H. M. (2020). A critical juncture in data protection standards: Comparing data protection legislation in the United States and the European Union. *Doctoral dissertation*, 서울대학교 대학원.

Czarnocki, J., Kun, E., Giglio, F., Royer, S., & Petik, M. (2021). Legal study on Government access to data in third countries. *Legal study on Government access to data in third countries.*

Diker Vanberg, A., & Maunick, M. (2018). Data protection in the UK post-Brexit: The only certainty is uncertainty. *International Review of Law, Computers & Technology, 32*(1), 190–206.

Dwivedi, Y. K., Hughes, D. L., Coombs, C., Constantiou, I., Duan, Y., Edwards, J. S., Gupta, B., Lal, B., Misra, S., Prashant, P., Raman, R., Rana, N. P., Sharma, S. K., & Upadhyay, N. (2020). Impact of COVID-19 pandemic on information management research and practice: Transforming education, work and life. *International Journal of Information Management, 55*, Article 102211.

Enright, H. (2022). What a data privacy law should look like in West Virginia: Balancing competing interests of consumers and businesses. *W. Va. L. Rev., 125*, 263.

European Union (EU). (2016). *General Data Protection Regulation (GDPR).* Retrieved from https://eur-lex.europa.eu/legal-content/EN/TXT/?uri=CELEX%3A32016R0679. Last accessed on December 20, 2024

Fahey, E. (2022). *The EU as a global digital actor: Institutionalising global data protection, trade, and cybersecurity.* Bloomsbury Publishing.

Ivanova, Y. (2021). The role of the EU Fundamental right to data protection in an algorithmic and big data world. *Data Protection and Privacy, Volume 13: Data Protection and Artificial Intelligence 13*, 145.

Justice K. S. Puttaswamy (Retd.) v. Union of India, (2017). Supreme Court of India. Retrieved from https://www.scobserver.in/cases/puttaswamy-v-union-of-india-fundamental-right-to-privacy-case-background/. Last accessed on December 20, 2024

Kontargyris, X. (2018). *IT laws in the era of cloud-computing: A comparative analysis between EU and US law on the case study of data protection and privacy.* Nomos Verlagsgesellschaft mbH & Co.

Kuner, C., Bygrave, L. A., Docksey, C., Drechsler, L., & Tosoni, L. (2021). *The EU general data protection regulation: A commentary/update of selected articles.* Update of Selected Articles (May 4, 2021).

Li, Z. S., Werner, C., Ernst, N., & Damian, D. (2022). Towards privacy compliance: A design science study in a small organization. *Information and Software Technology, 146*, Article 106868.

Lynskey, O. (2018). *The age of GDPR: The future of data protection law.* Cambridge University Press.

Ministry of Electronics and Information Technology. (2021). *The information technology rule, 2021: Amendments and updates.* Government of India. Retrieved from https://www.meity.gov.in/writereaddata/files/Revised-IT-Rules-2021-proposed-amended.pdf. Last accessed on December 20, 2024

Monahan, T. (2016). Built to lie: Investigating technologies of deception, surveillance, and control. *The Information Society, 32*(4), 229–240.

Motiwala, F. (2024). Comparative Study of General Data Protection Regulation (GDPR) and Digital Personal Data Protection Act (DPDP): A Way Out for European Industry to Navigate through Indian Personal Data Protection Regulation.

NASSCOM. (2023). *Impact assessment of the DPDP Act on the Indian IT industry.* National Association of Software and Service Companies. Retrieved from https://nasscom.in/. Last accessed on December 20, 2024

Reis, O., Eneh, N. E., Ehimuan, B., Anyanwu, A., Olorunsogo, T., & Abrahams, T. O. (2024). Privacy law challenges in the digital age: A global review of legislation and enforcement. *International Journal of Applied Research in Social Sciences, 6*(1), 73–88.

Singh, S., Singh, J., Goyal, S. B., El Barachi, M., & Kumar, M. (2023). Analytical review of map matching algorithms: Analysing the performance and efficiency using road dataset of the Indian subcontinent. *Archives of Computational Methods in Engineering, 30*(8), 4897–4916.

Sundara, K., & Narendran, N. (2023). The digital personal data protection act, 2023: Analysing India's dynamic approach to data protection. *Computer Law Review International, 24*(5), 129–141.

The Economist. (2017, May 6). The world's most valuable resource is no longer oil, but data. Retrieved from https://www.economist.com/leaders/2017/05/06/the-worlds-most-valuable-resource-is-no-longer-oil-but-data

Wells, C. P., & Ito, N. (2020). Personal information protection in Japan. *Journal of Data Protection & Privacy, 4*(1), 59–71.

Public Interest Litigation on Food Security and the Right to Livelihood

Insha Goel Vats

Abstract This study examines the pivotal role of Public Interest Litigations (PILs) in advancing food security and the right to livelihood in India, particularly through judicial activism. By exploring landmark cases such as PUCL v. Union of India (2001) and Olga Tellis v. Bombay Municipal Corporation (1985), the research highlights the judiciary's influence in transforming socioeconomic rights into enforceable legal entitlements. Employing a doctrinal methodology, the study analyzes judgments, statutes, and welfare policies like the National Food Security Act (NFSA) to assess their effectiveness. Findings reveal that judicial interventions have catalyzed policy reforms and improved governance in schemes like the Public Distribution System (PDS) and Mid-Day Meal Scheme. However, systemic inefficiencies, corruption, and limited beneficiary awareness remain persistent challenges. The study underscores the importance of collaborative efforts between the judiciary, executive, and civil society to bridge implementation gaps. While emphasizing the judiciary's transformative potential, it also identifies limitations in translating judicial mandates into tangible social outcomes. The research offers a comprehensive analysis of PILs' social implications, contributing original insights into the interplay between judicial activism, governance, and human rights.

Keywords Public interest litigation · Food security · Judicial activism · Right to livelihood · National food security act · Governance · Socioeconomic rights

1 Introduction

In India, public interest suits, or PILs, have long been a mainstay of judicial activism and serve as a crucial channel for addressing structural injustices. Over the years, the judiciary has used PILs to elevate marginalized populations' voices and compel the state to account for critical socioeconomic challenges. Among these, food security

I. G. Vats (✉)
Vivekananda Institute of Professional Studies - Technical Campus, New Delhi, India
e-mail: insha.vats@vips.edu

R. S. Mor et al. (eds.), *Sustainable Business Management, Innovation and Technology*, Sustainable Economy and Ecotechnology,
https://doi.org/10.1007/978-3-031-98141-8_19

and the right to livelihood stand out as critical areas where PILs have facilitated dramatic change. Despite constitutional provisions, hunger, malnutrition, and livelihood insecurity remain long-standing concerns in India's socioeconomic environment. PILs have addressed these issues by emphasizing the inseparable connection between dignity and nourishment as well as reaffirming the fundamental right to life under Article 21. This study investigates how PILs have impacted the legal and policy frameworks governing food security and the right to livelihood, examining their influence through landmark cases and judicial interventions. The analysis will concentrate on significant decisions that have influenced legislative and administrative reforms, looking at their consequences for governance and social justice.

2 Literature Review

Food Security and judicial activism in India[1]: With a focus on the justiciable aspect of the right to food, this book offers a thorough examination of judicial action in maintaining food security in India. The Public Distribution System (PDS) and other social programs' role in tackling food insecurity is discussed in detail in the book, which also highlights implementation issues like corruption and inefficiency. By portraying hunger as a breach of the fundamental right to life, the story links food security to human rights.

Parikshet Sirohi, *"Right to Food in India"*, Delhi Journal of Contemporary Law[2]: The legal and human rights aspects of India's right to food are examined in Dr. Parikshet Sirohi's paper. Citing its acceptance in international agreements such as the ICESCR and CEDAW, it explores how this right is integrated into the larger framework of human rights. It applauds the enactment of the NFSA as a milestone while discussing its limitations in addressing deep-rooted structural issues such as poverty and inequality.

Food Security in India—NCERT Analysis (2024–25 Edition)[3]: This chapter discusses India's strides in achieving food sufficiency since the Green Revolution and the establishment of welfare systems like the PDS. However, it critically evaluates the gaps in food distribution, which lead to persistent hunger and malnutrition among marginalized populations. The document highlights the role of the NFSA in expanding food entitlements and addresses contemporary issues, including the impact of natural disasters and pandemics like COVID-19 on food security. Through statistical data, it demonstrates progress in reducing hunger while recognizing ongoing challenges in ensuring equitable access to food for vulnerable groups.

---

[1] Ziegler, J. (2007).

[2] Sirohi, P. (n.d.).

[3] NCERT. (2024–2025).

3 Research Questions

1. How have Public Interest Litigations shaped the judicial interpretation and enforcement of food security as a fundamental right in India?
2. What role has judicial activism played in bridging the gaps in governance and policy implementation regarding the right to livelihood and food security?
3. How have landmark PIL cases influenced the development of welfare schemes like the National Food Security Act (NFSA)?

4 Research Methodology

This research adopts a qualitative approach, focusing on doctrinal analysis of legal judgments, statutes, and policy frameworks related to food security and livelihood rights in India. Primary data sources include landmark PIL judgments and subsequent judicial directives. Secondary sources, such as academic articles, government reports, and NGO publications, provide context and critique of judicial interventions.

4.1 Food Security and India

The concepts of food security and the right to livelihood have evolved significantly in India, shaped by the socio-political and economic conditions of different eras. In pre-independence India, famines were frequent and devastating due to colonial policies prioritizing resource extraction over local welfare. About three million people died during the Bengal Famine in 1943, underscoring the urgent need for structural changes to provide food security and sufficient livelihoods.[4] India had to deal with both economic instability and food scarcity after gaining independence. In response, the government took a number of actions to guarantee food security and reduce poverty.[5] A major turning point was the Green Revolution in the 1960s, which increased agricultural production, particularly for staple crops like rice and wheat. Even if food grains were self-sufficient, problems with pricing and accessibility remained, leaving a sizable portion of the populace food insecure. Through its Fundamental Rights and DPSPs, the Indian Constitution offers a strong basis for livelihood rights and food security. These provisions collectively mandate the state to assure equitable distribution of resources and access to basic necessities.

Article 21—Right to Life: The rights to life and personal freedom are protected under Article 21. Judicial interpretation has broadened the definition of this right over time to encompass the rights to livelihood, food, and housing. The Supreme

[4] Choudhury, S. (2021).

[5] Dwivedi, S. (2021).

Court ruled in seminal judgments such as *PUCL v. Union of India* (2001), [6] that the right to life includes the right to live with dignity, which is impossible to attain without access to sufficient food and a means of subsistence. The court's involvement in instances like *Olga Tellis v. Bombay Municipal Corporation* (1985)[7] highlighted the close connection between the right to life and the right to livelihood. The ruling ensured constitutional protection against capricious acts that endanger survival by establishing that stripping people of their livelihood without a proper procedure breaches Article 21.

Article 39(a)—Right to an Adequate Means of Livelihood: The state is required by Article 39(a) of the DPSPs to guarantee people' right to a sufficient standard of living. This clause emphasizes the significance of fair resource distribution and access to employment opportunities, despite the fact that it is not subject to judicial review. It establishes the framework for government programs like the Mahatma Gandhi National Rural Employment Guarantee Act (MGNREGA), which aims to give underprivileged populations jobs and improve their living stability.

Article 47—Duty of the State to Raise Nutrition Levels and Standards of Living: Article 47 places an obligation on the state to improve public health, nutrition, and living standards. This directive principle has guided the formulation of welfare programs such as the Public Distribution System (PDS), the Mid-Day Meal Scheme, and the Integrated Child Development Services (ICDS). These programs aim to address hunger and malnutrition, particularly among vulnerable groups like children and women.

The interplay between these constitutional provisions establishes a comprehensive framework that mandates the state to prioritize food security and livelihood rights as essential elements of social welfare and justice.

4.2 The National Food Security Act (NFSA) and Its Implications

An important turning point in India's food security legislation was the passage of the National Food Security Act (NFSA) in 2013. The Act aims to make the right to food a legally binding obligation by giving subsidized food grains to roughly two-thirds of India's population.

Key Features of the NFSA.

Coverage and Entitlements: "Up to 75% of people living in rural areas and 50% of people living in urban areas are covered by the NFSA. Five kilos of food grains per person per month at heavily subsidized rates (Rs. 3 per kg for rice, Rs. 2 for wheat, and Rs. 1 for coarse grains) are granted to priority households."

---

[6] People's Union Of Civil Liberties v. Union of India, AIR 1997 SC 568.

[7] Olga Tellis v. Bombay Municipal Corporation, 1985 SCC (3) 545.

Antyodaya Anna Yojana (AAY): The Act provides for the most vulnerable of the poor, guaranteeing that each household receives 35 kg of food grains each month.

Nutritional Support: Free meals are provided through programs like the Mid-Day Meal Program and ICDS, and special accommodations are made for expectant mothers, nursing mothers, and toddlers.

Grievance Redressal Mechanism: To tackle enforcement and access difficulties, the Act requires the creation of grievance redressal authorities at the state and district levels.

Implementation Challenges: Despite its transformative potential, the National Food Security Act (NFSA) faces significant implementation challenges, including persistent issues in accurately identifying beneficiaries, leading to errors of exclusion and inclusion that undermine its effectiveness. Furthermore, inadequate storage facilities and inefficiencies in the supply chain result in food wastage and delays in distribution, while corruption and leakages in the Public Distribution System (PDS), a key mechanism for implementing the Act, further compromise its objectives by enabling the diversion of food grains meant for the needy.

Impact on Vulnerable Groups: Hunger and malnutrition have decreased as a result of the NFSA's substantial improvements to underserved populations' access to food. The Act's programs, such the Mid-Day Meal Scheme, have improved children's nutritional outcomes and increased school attendance. However, persistent inequalities and regional disparities necessitate continuous efforts to strengthen the Act's implementation.

4.3 International Commitments

India's commitment to food security and livelihood rights is reinforced by its duty under international law. These commitments provide a broader framework for understanding and addressing the issues of hunger and livelihood insecurity.

International Covenant on Economic, Social and Cultural Rights (ICESCR)

The right to a sufficient standard of life, which includes enough food, clothing, and housing, is recognized by Article 11 of the ICESCR. The fundamental right to be free from hunger is also emphasized.

After ratifying the ICESCR, "India is required to take proactive measures to guarantee that its people have access to enough food that is safe and nourishing." Domestic policy such as the NFSA and other welfare programs have been impacted by this dedication.

Universal Declaration of Human Rights (UDHR)

"Everybody has the right to a standard of living that is sufficient for their health and well-being, including access to food, clothing, shelter, and medical treatment, according to Article 25 of the UDHR."

Although non-binding, the UDHR serves as a foundational document that underscores the universality of food security and livelihood rights, guiding national and international efforts to address hunger and poverty.

Sustainable Development Goals (SDGs)

Goal 2 of the SDGs aims to achieve zero hunger by 2030, emphasizing the need for sustainable food systems and improved nutrition.

India's participation in the SDG framework aligns with its domestic efforts to address food security and promote sustainable livelihoods.

The legal framework for food security and livelihood rights in India is deeply rooted in constitutional mandates, national legislation, and international commitments. While significant progress has been made, challenges such as systemic inefficiencies, corruption, and regional disparities persist. The interplay between judicial activism, policy initiatives like the NFSA, and international obligations underscores the multidimensional nature of addressing food security and livelihood issues. Moving forward, a concerted effort is required to strengthen implementation mechanisms, ensure accountability, and bridge gaps in access to food and livelihood resources for all citizens.

4.4 The Role of PIL

Public Interest Litigations (PILs) in India have served as a vital tool for addressing systemic gaps in governance and ensuring the realization of fundamental rights. In the domain of food security and livelihood rights, the judiciary has played a transformative role, particularly through landmark cases that have expanded the scope of Articles 21 and 39(a) of the Constitution. These cases are not isolated incidents but part of a larger narrative, where each builds on the other to create a coherent framework for securing these rights.

The case of "*People's Union for Civil Liberties (PUCL) v. Union of India*" (2001) stands as a cornerstone in the history of judicial intervention in food security. Filed during a period of severe starvation deaths in Rajasthan, the PIL underscored the stark contrast between surplus food grains in government warehouses and the hunger experienced by millions. By directing state governments to convert existing food security schemes into enforceable legal entitlements, "the Supreme Court ensured that programs like the Public Distribution System (PDS), Integrated Child Development Services (ICDS), and the Mid-Day Meal Scheme were transformed into lifelines for vulnerable populations." The court's appointment of food commissioners and its interim orders brought systemic inefficiencies to the forefront, laying the groundwork for the enactment of the "National Food Security Act (NFSA)" in 2013.

This foundational case set the tone for subsequent judicial interventions, such as *Kishen Pattnayak v. State of Orissa* (1989).[8] Here, the Supreme Court addressed starvation deaths caused by prolonged drought in Odisha. Recognizing the negligence of state authorities, the court mandated the establishment of food relief centers and the timely delivery of food grains to affected regions. This case emphasized the state's proactive obligation to ensure food security during emergencies, further enriching the jurisprudence initiated by the PUCL case.

Expanding the boundaries of Article 21, *Sunil Batra v. Delhi Administration* (1978)[9] highlighted the human rights of prisoners. Although primarily focused on the inhumane conditions in prisons, the judgment affirmed that access to adequate food and nutrition is a fundamental necessity tied to dignity and life. This interpretation seamlessly connects with "*Olga Tellis v. Bombay Municipal Corporation* (1985), where the Supreme Court linked the right to livelihood with the right to life." By protecting the rights of pavement dwellers against eviction, the court acknowledged that without livelihood, access to food and basic necessities becomes unattainable, strengthening the interdependence of these rights.

Further reinforcing this principle, "*Chameli Singh v. State of Uttar Pradesh* (1996)"[10] expanded the scope of the right to life to include food, clothing, and a clean environment. The court's emphasis on ensuring basic necessities for a dignified existence echoes through "*Paschim Banga Khet Mazdoor Samity v. State of West Bengal* (1996)".[11] While this case focused on inadequate healthcare facilities, the court underscored the importance of nutrition as a social determinant of health, thereby integrating food security into the broader framework of public health rights. The judiciary's proactive stance is also evident in cases addressing systemic barriers in public health and sanitation. For instance, *Municipal Council, Ratlam v. Vardichan* (1980)[12] directed municipal authorities to address public health hazards, indirectly reinforcing the importance of addressing malnutrition and food insecurity. Similarly, the Delhi High Court's ruling in *Harit Recyclers Association v. Union of India* (2010)[13] addressed the poor quality of food served under the Mid-Day Meal Scheme. The court's directive to implement stricter quality controls significantly improved the program, benefiting millions of children.

Building on these judicial precedents, cases like *M.C. Mehta v. Union of India* (2001)[14] emphasized the need for cooked meals under the Mid-Day Meal Scheme, enhancing its impact on child nutrition. The Delhi High Court's judgment in "*Laxmi*

---

[8] Kishen Pattnayak v. State of Orissa, AIR 1989 SC 677.

[9] Sunil Batra v. Delhi Administration, 1978 SCC (4) 494.

[10] Chameli Singh v. State of Uttar Pradesh, AIR 1996 SC 1051.

[11] Paschim Banga Khet Mazdoor Samity v. State of West Bengal, (1996) 4 SCC 37.

[12] Municipal Council, Ratlam v. Vardichan, 1980 SCC (4) 162.

[13] Harit Recyclers Association v. Union of India, WP (C) No. 2714 of 2010.

[14] M.C. Mehta v. Union of India, AIR 2001 SC 1848.

Mandal v. Deen Dayal Harinagar Hospital (2010)"[15] took this further by linking maternal health to food security, directing the better implementation of ICDS to support pregnant women and children. The role of courts in strengthening the Public Distribution System (PDS) has been particularly noteworthy. In the aftermath of the PUCL case, the Supreme Court's monitoring committees worked to ensure transparency and reduce corruption in PDS operations. Technological reforms like Aadhaar-based authentication systems have since improved efficiency and minimized leakages. However, challenges persist, including the exclusion of eligible beneficiaries and delays in the distribution of food grains.

Judicial interventions have also significantly influenced governance and policy-making. For example, the Mid-Day Meal Scheme, which now benefits over 120 million children annually, owes much of its expansion and improvement to court directives. Similarly, court orders have ensured timely delivery of food and supplements under ICDS, reducing malnutrition and infant mortality rates. "The Mahatma Gandhi National Rural Employment Guarantee Act (MGNREGA) has also benefited from judicial scrutiny", which addressed wage delays and improved implementation, thereby reinforcing rural livelihood security.

The interconnectedness of these cases illustrates the judiciary's evolving approach to food security and livelihood rights. Each judgment builds on the principles established in earlier cases, creating a cohesive narrative that underscores the importance of these rights in ensuring a dignified existence (Fig. 1).

The graph depicts the impact of Public Interest Litigation (PIL) on food security in India:

The poverty rate line shows a decline from 26% in 2001 to 8% in 2020, reflecting the impact of economic growth, government welfare programs, and targeted food security measures like the Public Distribution System (PDS) and NFSA 2013. The malnutrition rate line shows that the child malnutrition (percentage of underweight children under 5 years) dropped from 48% in 2001 to 32% in 2020, indicating improved nutrition access due to judicial interventions like the PUCL case and schemes like the mid-day meal program.[16]

5 Conclusion and Recommendation

Landmark cases have not only strengthened the link between sustenance and dignity but also catalyzed legislative milestones like the National Food Security Act. However, challenges such as systemic inefficiencies, corruption, and gaps in awareness and implementation persist, limiting the judiciary's transformative potential. To address these shortcomings and ensure the full realization of food security and livelihood rights, the following recommendations are proposed:

[15] Laxmi Mandal v. Deen Dayal Harinagar Hospital, W.P.(C) 8853/2008.

[16] NCERT. (2024–2025).

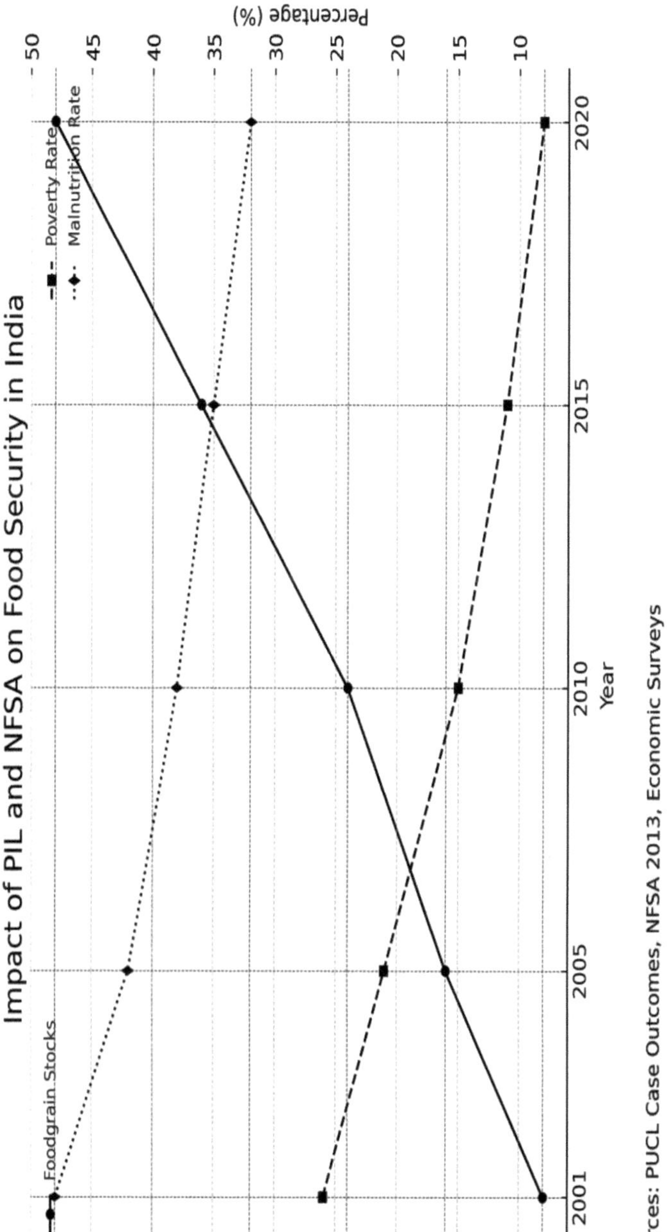

Fig. 1 Impact of PIL and NFSA on food security in India

Robust mechanisms must be developed to improve the implementation of judicial directives and welfare schemes. This includes investing in infrastructure for food storage, streamlining supply chains, and leveraging technology to reduce leakages and delays.

Governance reforms should focus on enhancing transparency in food security programs. Digital tools such as Aadhaar-based authentication systems and real-time monitoring should be utilized to track beneficiaries, reduce corruption, and ensure accountability.

State and district-level authorities responsible for implementing food security schemes require adequate training and resources to handle logistical and administrative challenges effectively. Regular audits and performance reviews can help improve efficiency.

Awareness campaigns must be conducted to educate beneficiaries about their entitlements under laws like the NFSA.

Targeted interventions are needed to address regional inequalities and cater to the specific needs of vulnerable groups, such as women, children, and tribal populations. Integrating food security with public health and employment programs can offer holistic solutions to interconnected challenges.

Innovations in agriculture, such as promoting climate-resilient crops, can also enhance food security sustainably.

The judiciary has been a pivotal actor in addressing systemic failures and ensuring that food security and livelihood rights are treated as non-negotiable components of social justice. However, judicial interventions are not a panacea. A collaborative effort involving the judiciary, executive, civil society, and local communities is essential to overcome implementation challenges and ensure that no individual suffers from hunger or livelihood insecurity. By bridging the gap between policy intentions and on-ground realities, India can move closer to achieving its constitutional promise of a dignified life for all its citizens.

References

Choudhury, S. (2021). The Bengal famine of 1943: Misfortune or imperial schema. *Cognizance Journal of Multidisciplinary Studies.* https://doi.org/10.47760/cognizance.2021.v01i05.002

Dwivedi, S. (2021). Raising productivity of cereal crops in dry and heat stress environments remains a breeding challenge. *Cab Reviews: Perspectives in Agriculture, Veterinary Science, Nutrition and Natural Resources 16.* https://doi.org/10.1079/PAVSNNR202116008

Harit Recyclers Association v. Union of India, WP (C) No. 2714 of 2010.

Kishen Pattnayak v. State of Orissa, AIR 1989 SC 677.

Laxmi Mandal v. Deen Dayal Harinagar Hospital, W.P.(C) 8853/2008.

M.C. Mehta v. Union of India, AIR 2001 SC 1848.

Municipal Council, Ratlam v. Vardichan. (1980). SCC (4) 162.

NCERT. (2024–2025). *Food security in India.* National Council of Educational Research and Training. https://ncert.nic.in/textbook/pdf/iess204.pdf

Olga Tellis v. Bombay Municipal Corporation. (1985). SCC (3) 545.

Paschim Banga Khet Mazdoor Samity v. State of West Bengal. (1996). 4 SCC 37.

People's Union of Civil Liberties v. Union of India, AIR 1997 SC 568.
Sunil Batra v. Delhi Administration. (1978). SCC (4) 494.
Sirohi, P. (n.d.). Right to food in India. *Delhi Journal of Contemporary Law 4*. e-ISSN 2582-4570.
Ziegler, J. (2007). *Food security and judicial activism in India*. Human Rights Law Network.